Biology, Ethics, and the Origins of Life

THE JONES AND BARTLETT SERIES IN PHILOSOPHY
Robert Ginsberg, General Editor

Ayer, A. J., *Metaphysics and Common Sense*, 1994 reprint with corrections and new introduction by Thomas Magnell, Drew University

Beckwith, Francis J., University of Nevada, Las Vegas, Editor, *Do the Right Thing: A Philosophical Dialogue on the Moral and Social Issues of Our Time*

Caws, Peter, George Washington University, *Ethics from Experience*

Cooper, Jane, The Pennsylvania State University, Delaware County Campus, *A Comparative Primer for Ethics: With Applications in Biology and Medicine*

DeMarco, Joseph P., Cleveland State University, *Moral Theory: A Contemporary View*

Devine, Philip E., Providence College, and Celia Wolf-Devine, Stonehill College, Editors, *Sex and Gender*

Edgar, Stacey L., State University of New York, Geneseo, *Computer Ethics*

Gorr, Michael, Illinios State University, and Sterling Harwood, San Jose State University, Editors, *Crime and Punishment: Philosophic Explorations*

Harwood, Sterling, San Jose State University, Editor, *Business as Ethical and Business as Usual, Text and Readings*

Heil, John, Davidson College, *First-Order Logic: A Concise Introduction*

Jason, Gary, San Diego State University, *Introduction to Logic*

Jason, Gary, San Diego State University, *Critical Thinking: Plain Speaking and Clear Understanding*

Merrill, Sarah A., Purdue University/Calumet, *Ethical Challenges in Construction, Engineering and Contracting, Text and Case Studies*

Moriarty, Marilyn, Hollins College, *The Conceptual Handbook to Scientific Writing: Critical Thinking Through Writing*

Pauling, Linus and Daisaku Ikeda, *A Lifelong Quest for Peace, A Dialogue*, Translator and Editor, Richard L. Gage

Pojman, Louis P., The University of Mississippi and Francis J. Beckwith, University of Nevada, Las Vegas, Editors, *The Abortion Controversy: A Reader*

Pojman, Louis P., The University of Mississippi, Editor, *Environmental Ethics: Readings in Theory and Application*

Pojman, Louis P., The University of Mississippi, *Life and Death: Grappling with the Moral Dilemmas of Our Time*

Pojman, Louis P., The University of Mississippi, Editor, *Life and Death: A Reader in Moral Problems*

Rolston, Holmes, III, Colorado State University, Editor, *Biology, Ethics, and the Origins of Life*

Townsend, Dabney, The University of Texas at Arlington, Editor, *Readings in Aesthetics*

Veatch, Robert M., The Kennedy Institute of Ethics, Georgetown University, Editor, *Cross-Cultural Perspectives in Medical Ethics: Readings*

Veatch, Robert M., The Kennedy Institute of Ethics, Georgetown University, Editor, *Medical Ethics*

Verene, Donald P., Emory University, Editor, *Sexual Love and Western Morality: A Philosophical Anthology*

Williams, Clifford, Trinity College, Illinois, Editor, *On Love and Friendship: Philosophical Readings*

Biology, Ethics, and the Origins of Life

Edited by

HOLMES ROLSTON, III

JONES AND BARTLETT PUBLISHERS
BOSTON   LONDON

Editorial, Sales, and Customer Service Offices
Jones and Bartlett Publishers
One Exeter Plaza
Boston, MA 02116
1-800-832-0034
1-617-859-3900

Jones and Bartlett Publishers International
7 Melrose Terrace
London W6 7RL
England

Library of Congress Cataloging-in-Publication Data
Biology, ethics, and the origins of life / Holmes Rolston, III, editor.
    p. cm.
    Results of a conference held at Colorado State University, Sept.
26–29, 1991.
    Includes bibliographical references and index.
    ISBN 0–86720–875–9
    1. Human ecology—Moral and ethical aspects—Congresses.
    2. Evolution (Biology)—Moral and ethical aspects—Congresses.
    3. Biology—Philosophy—Congresses. 4. Life—Origin—Congresses.
    I. Rolston, Holmes
    BD581.B52   1995
    171'.7—dc20   94–26715
                                         CIP

Acquisitions Editors: Arthur C. Bartlett and Nancy E. Bartlett
Production Editor: Anne S. Noonan
Manufacturing Buyer: Dana L. Cerrito
Design: Laura L. Cleveland, WordCrafters Editorial Services, Inc.
Editorial Production Service: WordCrafters Editorial Services, Inc.
Illustrations: Freehold Studio
Typesetting: A & B Typesetters, Inc.
Cover Design: Hannus Design Associates
Printing and Binding: Braun-Brumfield, Inc.
Cover Printing: New England Book Components

Printed in the United States of America
98 97 96 95 94   10 9 8 7 6 5 4 3 2 1

Contents

Biology, Ethics, and the Origins of Life

An Introduction

HOLMES ROLSTON, III

■ ■ ■ ─────────────────────────────────

Two critical points are of intense biological and philosophical interest in the natural history of life on Earth. The first is the origin of life; the second the origin of human life. With the first, biology began; with the second, ethics began. In the history of life on Earth, the chemical evolution of life is the opening chapter and the anthropoid evolution of human life is the latest chapter, but a final chapter that opens up a whole new drama. And in this story of human life, we today stand at a crossroads in such way that the Earth's history is at that junction with us. The chapters in this anthology move from the beginning to the present, trying to understand ethics in its interaction with biology, evaluating origins to discover the nature in, and the nature of, our duties.

Although life has continued on Earth for several billion years and human life has continued for a hundred thousand years, there is a profound sense in which we humans in the twentieth century, in an age of science, turning the next millennium, know for the first time who we are and where we are. This has been Darwin's century, and we have more understanding than any people before us of the evolutionary natural history by which we arrived on this planet. We know the astronomical prehistory of Earth; we know its geological history. We know the natural history that lies behind us, around us, and out of which we have come.

With this knowledge comes power, for the science through which we understand the world also gives us power to change it. More than any people before us, owing to our technological prowess, we humans today have the capacity to do good and evil, to make war or to feed others, to act in justice and in love. Nor is it only the human fate that lies in our hands. We are altering the natural history of the planet, threatening alike the future of life and of human life. With such increasing knowledge and power comes increasing duty. Science demands conscience.

Yet, although there is a profound sense in which we now know who and where we are, there is an equally deep puzzlement about the grounds of what we ought to do. Science has made us increasingly competent in knowledge and power, but it has also made us decreasingly confident about right and wrong. The evolutionary past has not been easy to connect with the ethical future. To the contrary, there is a widespread conviction that this cannot be done, that to do so is to commit the naturalistic fallacy, moving without warrant from what *is* the case to what *ought* to be. There is no route from biology to ethics despite the fact that here we are, *Homo sapiens*, the wise species, lacking wisdom, troubled by ethical concerns, with our own future and that of our planet in our hands.

These three features of our human life on Earth—knowledge, power, and duty—are especially philosophically puzzling. How does reason, the mind with its knowledge, fit into the biological picture? Does it produce only more survival power? Does not this human mind gain some new power that changes the evolutionary story? Such knowledge and power, descriptive facts of modern life, seem inescapably to prescribe an ought. But the same science that demands a conscience has difficulty explaining and authorizing conscience, for we struggle to understand how amoral nature evolved the moral animal, how even now *Homo sapiens* has duties, humans to fellow humans, and humans to the community of life on Earth. Though we know more than ever before about who and where we are, the value questions in the twentieth century have been as sharp and as painful as ever in our history.

The origin of life and the origin of human life are events widely separated in time, perhaps two billion years apart and disparate in degree. One assembled the simplest kind of life, replicating molecules; the other, the most complex; the one an event in spontaneous nature, the other launching culture; the one an objective event, the other producing self-conscious subjects. But relate the two we must, if our analysis is, as Plato once put it, to carve nature at the joints (*Phaedrus* 265e). These are vital articulations, hinge points in the history of life. We are doubtfully competent to understand what is going on across the evolutionary epochs if we do not understand either the primordial beginning or this contemporary end.

These two joints are events of great discovery and moment; no other events on Earth are more challenging to fit into the big philosophical picture. At both these events, though widely separated, biology touches metaphysics. In the opening chapter and again in the human chapter, the story of Earth takes a quantum leap. When out of nonliving nature there arises living nature, when out of amoral nature there arises moral human life, the conclusions seem to exceed the premises; there is more out of less. We struggle to understand how and why this has taken place, and what is the meaning of this Earth story in which, willing or not, we humans now stand at a crossroads of opportunity.

At the origin of life, at the origin of human and ethical life, it can seem that possibilities float in from nowhere, to borrow Whitehead's way of phrasing it (recalled by Langdon Gilkey, Chapter 7)—an appearance that troubles several of our authors. When B follows A, do we interpret B in terms of A, or A in terms of B? Is B merely the unfolding result of A, or has A to be reinterpreted in the light of B, once potential within A and now become actual out of it? When biology chronologically follows geophysics and geochemistry, and life originates from nonbiotic materials, how do we understand these possibilities that become actual? When the social sciences follow biology, do we expect culture to be the outcome of nature, explained in natural categories, or do we see biology as necessary but not sufficient to explain the social sciences, perhaps something like physics is necessary but not sufficient to explain biological life? Where and how does social science pass over into philosophy and ethics? Eight eminent biologists, philosophers, and theologians explore these issues in the pages that follow.

The claims we just made about the extent of our knowledge notwithstanding, we are still struggling to understand the origins of life on Earth. In the opening chapter, Thomas R. Cech portrays the primordial world as an RNA world, where RNA assembles itself, combining both the informational and the functional roles necessary for life. These roles, in the subsequent story of life, have been played by DNA, the informational molecules, and proteins, the structural and functional molecules. His account is based on his own discovery of how RNA acts in both the coding and the metabolic roles of life, combined with the discovery by others of RNA in "molecular fossils" still imbedded deeply in the continuing life metabolisms of today. Perhaps it is too much to say that we yet know the whole story of the chemical incubation of life, but we now have a story plausible at least in its outlines.

Reflecting philosophically over his discoveries about the primordial biology, Cech refuses to devalue life in result. To the contrary, far from devaluing life, life is increasingly to be seen as a consequence of the earthen chemistries, and the possibilities and power of Earth are the more to be valued. "I feel much the other way about the origin of life, that this makes life even more special. If intrinsic to these small organic molecules is their propensity to self-assemble, leading to a series of events that cause life forms to originate, that is perhaps the highest form of creation that one could imagine" (Cech, Chapter 1, p. 33). Continuing, Cech insists that his account of the origins of life does not diminish our respect for life today, nor our ethical responsibilities for its protection.

Cech asks about life in the beginning and at the microscopic scale, but when we turn to Dorion Sagan and Lynn Margulis, the question is about life today at the global scale. For them, an explanation of life requires the Gaia hypothesis: Earth is a superorganism in which the myriad organisms on Earth are located. "The Gaia hypothesis says in essence that the entire Earth functions as a . . . responsive organism. . . . The planet [is] an amorphic, but viable biological entity" (Sagan and Margulis, 1984, p. 66, p. 69). Sagan and Margulis can amply emphasize life at the microbial levels, the oldest and longest continuing manifestation of Gaia, beside which human life is ephemeral. "Microbes, the first forms of life to evolve, seem in fact to be at the very center of the Gaian phenomenon. . . . Insofar as the larger forms of animal and plant life are essentially collections of interacting microbes, Gaia may be thought of as still a microbial phenomenon" (Sagan and Margulis, 1984, pp. 67–70).

But Sagan and Margulis wish to focus here on Gaia at the planetary level. A repeated truth is that life on Earth merges, and here Lynn Margulis invokes the discovery for which she is so justly celebrated, that synthesis and symbiosis have resulted in the mitochondria incorporated into and empowering eucaryotic cells, all our human cells included, as well as the chloroplasts incorporated into plants. She builds on the principle that life fuses repeatedly and from level to level to form recreated biological identities. This merging of life identities is not understood at the highest levels until it is understood at the systemic, planetary level. "Biological identity is not fixed. Identities that flow into each other require constant reaffirmation. Individuals join to produce new identities at more inclusive levels. This kind of merging to form larger individuals is the way of the world, and applies also to humans. . . . The line demarcating inclusion within an ethically protected community widens with time. The ethical frontiers bleed, expanding to the global community of life, the biosphere. Sooner or later, we face the Earth" (Sagan and Margulis, Chapter 2, pp. 60–61).

If so, then Earth as a responsive, superorganismic entity must be "faced," not only if we are to get our biology right, but if we are to get our philosophy and ethics right. Something appears on Earth that gives rise to a global sense of obligation. Indeed, such an Earth has a "face," a model or symbol by which Sagan and Margulis wish to identify this unique systemic presence of life on Earth. An appropriate Earth ethics can only occur in this facing relationship, which they describe as a kind of "epiphany," choosing a religious word for the appearance of something deeply revealing. That relationship will not emphasize the genetic selfishness of organisms (anticipating the authors to follow), nor will it suppose that Gaia is a moral agent. Nature, earthen nature, culminating in Gaia, is beyond good and evil. Earth is of fundamental value notwithstanding and demands our respect. Earth ethics, global scale, is the ethical frontier. "We should rejoice in the new truths of our essential belonging, our relative unimportance, and our complete dependence upon a biosphere which has always had a life entirely its own" (Sagan and Margulis, 1984, p. 73).

The originating of life is not, of course, all at the original start-up; life originates over the millennia, as novel kinds of life arise in the turnovers of species. Possibilities become actualities regularly when, over the ages, Earth's biological powers rise to new achievements. So the epochs of natural history between the critical first and the latest chapters are also relevant. Following Cech's account of the original replicating molecules at the microscopic scale and Sagan and Margulis's account of the face of the Earth at the global scale, we turn to Niles Eldredge's paleontological account on geological time scales. Science has made it possible for us to recall with increasing detail the evolutionary story, one that Eldredge here sketches.

But now a different aspect of the story strikes us. Cech's molecular life, intrinsic within these earthen molecules from the start, becomes afterward a prolific natural history marked by massive contingency. Sagan and Margulis's self-regulating superorganism seems to undergo convulsions. Eldredge is impressed with the repeated mass extinctions in the past; the forms of life that remain do so as much by luck as by natural selection. Eldredge recounts "the extremely checkered career that life has had on Earth" (Eldredge, Chapter 3, p. 70). If we were to "run the tape" over again, if (though it is impossible) the Earth story could happen again, it would be vastly different.

Against such contingency, Eldredge sets another fact, the resilience of life: "It has seemed to a number of biologists (particularly we paleobiologists) that Earth's biota is tough, able to rebound in both an evolutionary and ecological sense after even the worst of biotic devastations" (Eldredge, Chapter 3, p. 68). So although extinction comes to every life form, sometimes catastrophically, extinction is never the last word—at least it has never yet been. Eldredge finds that extinction is always coupled with respeciation, indeed it becomes a kind of key to Earth's most novel epochs of respeciation. There is a mixture of luck and toughness. We get, rather puzzlingly, a mixture of life persisting in the midst of its perpetual perishing. Eldredge accentuates at the same time the massive extinctions and the tough resilience of life. Sagan and Margulis will probably reply that we are picking up evidence that supports their Gaia hypothesis.

Those are the facts of the matter, the natural history Eldredge describes, which he recounts by way of anticipating his ethical concern: that the latecoming humans, especially we today, the latest of these humans, have responsibilities on this planet which has had such a striking history of life hanging tough against extinction. "Certainly, I subscribe to the view that we as individuals were born into, and as a collective species evolved in, a natural world that we ought to feel compelled to restore and protect" (Eldredge, Chapter 4, pp. 68–69). But Eldredge, a biologist, is wary about obtaining this ethics from his biology, however much he desires an ethics to protect the values that biology has produced on Earth. Since human life is inseparably entwined with the natural environment, he can urge his case for conservation in terms of our own human species self-interests, indeed of our own survival. That one ought to promote one's own welfare and survival is a conclusion familiar enough to biology. The resulting prescription—or, if ethicists prefer, prudential recommendation—is no less on a global scale than that of Sagan and Margulis: a concern for the entire community of life on Earth.

The authors in the second half of the book concern themselves more directly with the biological origin of ethics and the bearing of this on how ethics operates today. Michael Ruse, a philosopher, is much less reticent than Eldredge about deriving an ethics from biology. The ethics he enjoins, though, is equally as joined to self-interest as was that of Eldredge, indeed even more so. "What excites the evolutionist is the fact that we have feelings of moral obligation laid over our brute biological nature, inclining us to be decent for altruistic reasons" (Ruse, Chapter 4, p. 97). Such sentiments are genetically innate, and, as natural selection predicts, they serve our survival interests. Humans are by nature as cooperative a species as they are a combative species. "Especially between members of the same species, much more personal benefit can frequently be achieved through cooperation—a kind of enlightened self-interest" (Ruse, Chapter 4, p. 95).

Biological altruism has a technical, analogical, and nonmoral meaning: any behavior where one organism cooperates with another to serve its own survival interests. When humans arise, there appears genuine moral altruism, as one means by which biological altruism is achieved. "Literal, moral altruism is a major way in which advantageous biological cooperation is achieved. Humans are the kinds of animals that benefit biologically from cooperation within their groups (biological altruism), and (literal, moral) altruism is the way in which we achieve that end" (Ruse, Chapter 4, p. 96). The explanatory principle is that if morality, B, sequentially follows natural selection, A, we must interpret B in terms of A, a genetic explanation. Natural selection is the criterion against which ethical theories are to be measured.

There is a troublesome side to this derivation of ethics from biology. Biology may produce literal altruism, but it also produces illusory altruism. "In other words, we see that morality has no objective foundation. It is just an illusion, fobbed off on us to promote 'altruism'" (Ruse, Chapter 4, p. 100). Ethics is not true, though it is functional. Paradoxically, though, ethics cannot be functional unless it is believed to be true in an objective sense, a false belief. Ethics is, at best, true for us humans in our biological character; or, more frankly put, it is the most reasonable way to behave, if one wishes to survive and reproduce.

Ruse then struggles to make ethics, so much needed for human flourishing, an acceptable, though ultimately false, illusion. To this end, he seeks to show that his evolutionary ethics has close affinities with John Rawls's concept of justice, where enlightened and reflective persons cooperate in their own self-interests to form a just society. "If we take modern biology seriously, we . . . learn what the true situation really is. Evolution and ethics are at last united in a profitable symbiosis. . . . We have succeeded in a unified theory combining biology, ethics, and the origins of human life, a theory that can and ought to govern our contemporary and future practice" (Ruse, Chapter 4, p. 109).

While for Ruse ethics is fundamentally and literally a survival tool, Francisco J. Ayala finds a difference in being human. Interestingly now, when we consult a biologist renowned for his studies in genetics, there is reluctance to derive ethics from biology (as there was before in Eldredge). Ethical behavior is an evolutionary byproduct. In the hominoid line at least, evolution selects for intelligence, but, in so selecting, even though survival value is always a result of the kind of intelligence that is selected, natural selection becomes dissociated with much of what that intelligence is used for. Many products of intelligence are what Ayala calls "byproducts." Another way he puts this point is that intelligence *generically* is selected for, but that the *specific* activities of intelligence are not. "Ethical behavior came about in evolution not because it is adaptive in itself, but as a necessary consequence of man's eminent intellectual abilities, which are an attribute directly promoted by natural selection" (Ayala, Chapter 5, p. 118).

Our biology, carefully speaking, requires no ethics at all; but our biology requires a general intelligence that brings ethics inevitably in its train (though nevertheless a byproduct because it is not required). Ethics is a kind of logical implication from intelligence, but not a biological implication from survival. In this indirect way, ethics is a capacity provided by nature, but all the normative contents are provided within culture. Nature does keep a threshold norm: any viable ethics must be consistent with our human nature, and there must be a next generation. So the fact that we are ethical, descriptively, is innate in our biology, but epiphenomenally innate (a byproduct); and that permits the norms of ethics, prescriptively, to be supplied by culture and subject to reasonable discourse and evaluation.

If we tried to derive an ethics from biology, we would indeed fall into the naturalistic fallacy, but here we escape that fallacy. We derive ethics from biology only in generic capacity, and that as byproduct; the content with which we fill this capacity is a cultural creation. "The evaluation of moral codes or human actions must take into account biological knowledge, but biology is insufficient for determining which moral codes are, or should be, accepted. . . . Moral norms are not determined by biological processes, but by cultural traditions and principles that are products of human history. That is the difference of being human" (Ayala, Chapter 5, p. 134). Ayala concludes, contrary to Ruse, that such moral practices today may nor may not have survival value in maximizing offspring.

Sober examines what difference the human brain makes and finds that humans (again contrary to Ruse) can behave outside the determinants of natural selection. He puzzles over the human anomaly, and concludes:

> We now have an answer to the question of why the human brain can throw a monkey wrench into the idea that adaptationism applies to human behavior with the same force that it applies to behaviors in other species. The brain is a problem for adaptationism because the brain gives rise to a process that can oppose the process of biological selection. And it is not just that cultural selection can find itself at odds with biological selection. The important thing is that cultural selection can be more powerful than biological selection. . . . *Thoughts spread faster than human beings reproduce.* . . . A thought—even one that is neutral or deleterious with respect to my survival and reproduction—is something that may expand beyond the confines of the single brain that produced it. Ideas can plug into a network in which brains are linked to each other by relations of mutual influence. This is a confederation that our brains have effected (pp. 156–157).

Sober concludes that this permits human culture to transcend biological nature.

"Biological selection produced the brain, but the brain has set into motion a powerful process that can counteract the pressures of biological selection. The mind is more than a proximate mechanism for the behaviors that biological selection has favored. It is the basis of a selection process of its own, defined in terms of its own measures of fitness and heritability. Natural selection has given birth to a selection process that has floated free" (Sober, Chapter 6, p. 158). That, Sober thinks, opens up the possibility of ethics, parts of which may coincide with the requirements of natural selection, since human survival is, after all, a good thing, but other parts of which may float free from determination by natural selection. The change in ethical norms regarding slavery in the last century, for example, does not seem to have any explanation in evolutionary theory.

Langdon Gilkey, the theologian among our contributors, agrees with Ayala and Sober that there is a difference in being human. The challenge is to place that *humanum* (all the characteristics of human history, especially those that distinguish humans from biological natural history) in the story of life on Earth. In the sweep from the origin of life, to the origin of morality, to our present responsibilities, the beginning and middle of the story (the biology) will have to be interpreted in the light of the conclusions to which they lead (the *humanum*). If culture, B, must be interpreted as the outcome of biology, A, that is, B in terms of A, it is equally true that, for humans at least, biology must be interpreted as the precursor of culture, A in terms of B. "All of the facets of 'spirit' or 'reason,' the entire *humanum*, stretch back into the dimness and mystery of so-called matter, into the mystery of nature as the source and ground of all that we are. Here nature as 'matter'—nature as known by the physical sciences, as *reduced*, one might say—shows its deeper, and more mysterious, identity with nature as *also* the source of psyche and of spirit" (Gilkey, Chapter 7, p. 170).

Gilkey welcomes the incorporation of human history into natural history. But there is more. "We are in part self-directing, centered beings. . . . Strangely, we humans must choose, affirm, and use these possibilities of our nature. . . . The call to 'choose ourselves' . . . is one of the characteristics that makes us human; it is the source of responsibility, and so of the moral. . . . Morals and science alike spring from the *humanum*—the human world—what theology has classically termed 'the image of God.' " The difference in being human is that we have "spirit" and "freedom" (Gilkey, Chapter 7, p. 169). In the world picture into which biology must be fitted, "nature edges into mind; mind edges into nature" (Gilkey, Chapter 7, p. 168).

The sociobiological account of the origin of ethics does have some promise for placing ethics in the perspective of natural history. It sometimes shares with theology a realism about the selfish elements in human nature. "Sociobiologists are discovering a biological basis for what Christians have long called original sin" (Gilkey, Chapter 6, p. 181). But the deeper problem is that sociobiology, unwilling to place itself in the larger community of human disciplines, and attempting an account of human nature on its own, does not have any such unified account, but rather results in an incoherent dualism, illustrated in contradictions within the sociobiologists themselves.

Sociobiologists often have a higher morality than any to which they are entitled by the biological theories they hold. "Share what estimate of human sinfulness sociobiology may with theology, we are left within biology with the problem of obtaining a better morality. . . . The scope of the theory is thus radically limited" (Gilkey, Chapter 7, pp. 175–176). It cannot include the morality of the sociobiologists themselves. Liberal sociobiologists in fact learn their morality from their cultural traditions, not from their biology (agreeing here with Ayala), and they may even be forced to urge that culture educate within us an ethical sensitivity that transcends biology. Such ethical sensitivity has had its classical origins in the religious traditions distinctive to the various cultures.

Nor is this merely a problem for ethics; this is a problem for science itself, including biology. For their sociobiological theory supposes a level of deception in the pursuit of survival advantage that undermines the rational validity of science itself by the identical deceptive tendencies with which the science calls morality into question. We must be rationally free to be scientists evaluating the claims of biology about natural history; we must be morally free to be ethicists evaluating our responsibilities.

This higher morality evidences a possibility not yet accounted for by the biological theory. This possibility has become actual not only in such sociobiologists "enlightened" with an ethics that exceeds their biology (Gilkey, Chapter 7, p. 176), but one that has been both possible and actual in the struggle to be human over the many centuries of the *humanum*, going back at least as far as such figures as Buddha, Christ, and other enlightened moral teachers of human history. There is no particular cause to think that genuine concern about morality appeared for the first time in history with sociobiology.

To the contrary, Gilkey insists, no objective scientific inquiry, in itself, is competent to evaluate the subjective, experienced sense of responsibility. But this experience too is both possibility and actuality quite as much as is the empirical science of natural history. The presence of genuine morality has to be reckoned with, as does the presence of moral failure. Biology as a science falls short of ethics. The "selfishness" alleged of nonmoral organisms is only a confused analogy that fails entirely to reckon with this dimension of experienced moral responsibility, since we do not suppose that such animals have any experience of felt responsibility, or remorse at their failures. We are left with a biology unable to authorize either the rationality or the ethics that biologists—and all who seek to be human—need to be human.

Charles Birch closes the anthology by advocating a postmodern theism. Readers will be perhaps surprised to find that a biologist, internationally known for his work in ecosystem science, is here more outspoken than the theologian who immediately precedes him. Birch claims that a theistic view is the metaphysics most congenial with the accounts of biology, ethics, and the origins of life that his colleagues have been attempting to understand. Biologists may do without philosophy in the laboratory, but biologists in real life, who want to be philosophically intelligent, need an account that goes "from protons to people" (Birch, Chapter 8, p. 209), and biology will have to be fitted into a total picture.

Darwinism, or neo-Darwinism, is the paradigm for biology—random mutations and natural selection of the fittest—and Birch examines that paradigm with two criticisms, the one for biology, the other for theology. The one is that, in the legacy of Newton, biology has been too mechanistic, too reductionistic, and hence stumbles to explain the really creative occasions in natural history, such as the origin of life and the origin of human moral life. But neither can classical forms of supernaturalist theism, or deism, reckon with such creativity, since they have too much of "a doctrine of divine carpentry" (Birch, Chapter 8, p. 198) in their concept of creation—an external God working on inert materials. The matter of Earth is not inert but agitated and prolific.

We do have to take contingency seriously; we do not live in either a determinist or a predestinarian universe. Biology takes chance seriously, both microscopic chance and macroscopic, even global chance, and mixes it creatively with the discovery of order. If possibilities so steadily become actualities not only at the origin of life, not only at the origin of human life, but all along the way en route, "there must be something positive limiting chance and something more than mere matter in matter" (Chapter 8, p. 205). True, if we were to run the tape over, the story would not be the same. But, nevertheless, we do have this long-continuing struggle for life on Earth. "A full biological and philosophical account is going to have to include this remarkable power of life to persist and develop over the millennia, despite the chances of its destruction" (Chapter 8, p. 203). "In this more comprehensive account, God is the name of the persuasive influence" (Chapter 8, p. 206). In a way, God serves for Birch as the unifying metaphysical postulate, much like Gaia does for Sagan and Margulis, though the latter claim Gaia to be as much science as metaphysics. God for Birch, like Gaia for Margulis and Sagan, is the principle of synthesis.

Critics will no doubt swiftly object that there is too much speculation and wish to return to "harder" science. But if we return to paradigmatic Darwinism, can we explain either the beginning or the end with which we have been trying to cope? Or, for that matter, can we really explain the millennia of advancing evolutionary history in between? In an era of science, it will be protested, can we not show that these events are "natural"? Is the finding that something is natural not explanation enough? Biologists may believe so, and certainly Thomas Cech helps us see how life may have "naturally" evolved. Sociobiologists hold that natural selection is competent to explain the origins of human moral life. But then again, we may be left wondering whether these naturalistic explanations are entirely convincing. Is it a finished explanation of such startling events as the origin of life or the origin of ethics to find that an event is "natural"? Biologically we may describe what historically took place; philosophically we still have to muse over the meaning of this startling creativity possible in nature and become actual over natural history.

The paradigm of biology is natural selection, but, at the ranges at issue here, at the beginning and the latest developments in the Earth story, we have to ask about the limits of the paradigm, for, at the start, natural selection itself has to originate somehow, and, at the human coming, in moral rationality, natural selection may pass over into something else. The Darwinian paradigm, with natural selection at its center, does not explain the origin of natural selection. Indeed, as the authors here several times remind us, even over the long route from the origin of life to the human coming, natural selection does not guarantee any steady progress. Nor do any here, Michael Ruse excepted, accept natural selection as a complete explanation of the origin of ethics. If we ask about the origins of biology, if we ask about the origins of ethics, if we ask for norms toward human life, so distinctive on Earth, or toward the fauna and flora surrounding us, biology has to pass over into philosophy. Perhaps it even reaches into theology, as Charles Birch insists. Pending answers, we still must wonder who we are and where we are.

Yet there are some convictions we can hardly escape. This is the home planet. So far as we know, Earth is the only planet with an evolution in the biological sense, the only planet with an ecology, the only planet with habitat, the only planet with a "face," with somebody there, the only planet with culture, the only planet with either biology or philosophy, the only planet with an ethical animal, with a species, *Homo sapiens*, that strives to become wise. We may hope that we are not alone in this vast universe, but even if there is extraterrestrial life, this would not gainsay either our own unique history nor our present responsibilities on this planet. The unexamined life, said Socrates, is not worth living; what we add here is that life in an unexamined world is not worthy living either, since we do not know who were until we know where we are, nor what our duties are until we know what on Earth is most valuable.

Reference

Sagan, Dorion, and Lynn Margulis. 1984. "Gaia and philosophy." In Leroy S. Rouner, ed., *On Nature* (pp. 60–75). Notre Dame, IN: University of Notre Dame Press.

This anthology results from a conference, "Biology, Ethics, and the Origins of Life," held at Colorado State University, September 26–29, 1991. The conference was made possible by generous support from the Applied Philosophy Endowment, the College of Natural Sciences, the College of Arts, Humanities and Social Sciences, the Office of the Vice-President for Research, and the Graduate School. Holmes Rolston III, the editor, is University Distinguished Professor of Philosophy at Colorado State University. He is the author of *Science and Religion: A Critical Survey* (1987), *Philosophy Gone Wild* (1986), *Environmental Ethics: Duties to and Values in the Natural World* (1988), and *Conserving Natural Value* (1994).

The Origin of Life and the Value of Life

THOMAS R. CECH

▪ ▪ ▪ *Editor's Introduction*

Thomas R. Cech begins with an inquiry into the beginnings of life. Evidently, both life and mind are among the possibilities in our universe, on our Earth, for here we are; but can we give an account of how those possibilities became actual? Hopefully, it can be a scientific account; but then, too, depending somewhat on how the possibilities become actual, and where the possibilities come from, we may need also a philosophical account, a metaphysical account. Such an account will also, sooner or later, need to become evaluative. Life has to be defended and, more, life ought to be defended. There is something about respect for life that is vital, both in the biological and the philosophical senses. Such evaluation will generate an ethics for the moral species and also for the larger community of life on Earth.

Life requires, as Cech points out, information flow. The secret of life is storing and using vital information. In that sense, when biology arises on Earth, there arises something more than just the matter and energy with which the physical sciences are concerned. There is information superimposed on certain states of matter and energy. It is with the origin of such informational, replicating molecules that Cech is here concerned, maintaining that one kind of molecule, RNA, can both store, functionally use, and replicate such information.

How did such molecules originate? Cech sketches a picture of RNA by which, under the conditions on early Earth, RNA was self-assembling. Thus arise molecules that can discover and carry information, code the metabolisms of life, reproducing this coding from generation to generation. Cech believes that he and other biologists are on the verge of solving the major problem of how to get life started—how to construct an informational molecule and make it self-replicating. Life began in "the original RNA world, . . . a time when RNA was both informational molecule and biocatalyst" (p. 30). The earliest proto-life molecule was self-synthesizing; it contained both information and function and was autocatalytic. All the processes of life, now executed by diverse molecules of DNA and the myriad proteins, were, at the start, executed by the one molecule, RNA.

Such is Cech's story of possibilities becoming actualities. Is it the case that possibilities float in from nowhere? Cech thinks not, that the possibilities are already intrinsic to the material. "At least from the perspective of a biologist, I have given an account of how possibilities did, in times past, become actual. When this happened, life originated with impressive creativity, and it does not seem to me that possibilities floated in from nowhere; they were already present, intrinsic to the chemical materials" (p. 33).

Cech's account makes the origin of life less contingent, more written into the nature of the chemicals. Life is, we might say, more natural. At the origin of life the luck factor is reduced, even though (as Eldredge will remind us) the luck factor still haunts much of subsequent natural history. "Life is to be expected as a consequence of chemical principles" (p. 31). Thus life, rare in the universe, becomes highly probable on Earth. But Cech does not on this account devalue it. To the contrary, we ought all the more to respect the intrinsic life properties at work on the planet.

The questions that Cech's discoveries and reflections raise are as vital for philosophy as are the processes he describes vital for life. Can we now say that life arose naturally? Is life somehow seeded into the chemistry? Into the elements severally or into the Earth systemically? Can we now say that life is "nothing but" chemistry? What is the mixture of the necessary and the contingent, the inevitable and the probable? With the discovery of the chemical evolution of life, can we say that life has been "explained away"? Is there any need for further explanation when we discover how life is natural? What is the connection between discovering how life originated and our present evaluation of the worth of life and of our responsibilities toward conserving it? As Cech asks: If life is generated naturally, does that make it any less special? Does there not remain an impressive creativity? And an important duty?

Thomas R. Cech holds a joint appointment as Professor of Chemistry and Biochemistry and Professor of Molecular, Cellular, and Developmental Biology at the University of Colorado, Boulder. He was awarded the Nobel Prize in 1989 for his discovery that RNA molecules can act as enzymes, a discovery that has revised our understanding of the origins of life. That award followed two dozen earlier awards recognizing his work. Dr. Cech is a member of the National Academy of Sciences. He oversees a state-of-the-art $3.4 million research laboratory, the Howard Hughes Medical Institute, with a staff of twenty persons, and an annual budget of over $1 million. A further introduction to his discoveries can be found in the articles cited in his bibliography.

To ask how life originated and what kind of respect for life we ought to have mixes questions from biology and from philosophy. They are not unrelated questions, because how we value life could depend, to some extent, on our picture of how life originated. So I want to ask what that picture is, and how it has been revised by the discovery that RNA can act as an enzyme. After that, I will try to look at this origin of life more philosophically, even ethically, and ask what difference this makes, if any, in the way we should now value life. I will be briefer about that; when one is treading on thin ice with regard to one's training, it is safest to spend as little time out on the ice as possible.

First, I will describe something of the molecular basis of all life on Earth, which requires molecules to serve two functions: to carry information and to act as catalysts. More particularly, I will describe how one molecule, ribonucleic acid (RNA), has the inherent capacity to serve as both an informational molecule and a catalytic molecule.

Second, based on this biochemical knowledge I will develop a plausible scenario for the origin of life on Earth. How might self-replicating molecules have evolved, as a result of RNA being both an informational and a catalytic molecule?

Third, I will ask whether we can really say that we know how life originated on Earth. Will we ever be able to have much certainty in pronouncements about such ancient events? With what level of knowledge about the origin of life must we be satisfied? There is some evidence for my account in "molecular fossils," features present universally in organisms living today that are thought to reflect features of the earliest life. I find this evidence consistent with theories of an RNA world, although less than compelling.

Fourth, I will ask about some of the philosophical and ethical consequences of my scenario for the origin of life.

1. *RNA as an Enzyme*

The philosophers contributing to this volume are later going to worry about how possibilities become actualities. When life originated something became actual that before was possible, and when life continues today, there is always something possible becoming actualized in the living

18

organism. From the viewpoint of a biologist, this requires, in all living systems on this planet, a flow of information. In general, the information instructs a cell how to make a particular protein. The protein could be a structural protein, such as a component of hair or skin, or it could be an enzyme that serves as a catalyst of a cellular process.

In contemporary organisms, the information that tells the cell how to build any of those proteins is encoded in the double helix of deoxyribonucleic acid, DNA. That information is copied onto a molecule of ribonucleic acid, RNA. Then, during the process of protein synthesis, the information from the RNA determines the particular sequence of amino acids that makes up a polypeptide chain. When that chain is folded, it becomes a protein. Then that protein will have a function, becoming part of the cellular structure of a hair or the skin or performing some metabolic catalysis, speeding up the rate of a specific chemical reaction needed for life.

An everyday analogy might be helpful. Within the cell, the DNA is something like a gold seal, archival mastercopy of a videotape. From that single mastercopy multiple, inexpensive, but still informationally equivalent copies of the tape can be reproduced (the RNA). Playing one of those copies on the VCR is equivalent to the cellular process of translation, in which the RNA information is used to construct a protein. The final product, in the case of video processing, is the image that one sees on a television monitor. In the cellular process, the product is a particular protein, functioning in the cell. Francis Crick, who most eloquently described the coding process, called this the central dogma of molecular biology, and it is still a good description of information transfer in living cells.

Until about ten years ago, it seemed that there was an almost complete division of labor between the molecules that store and carry the information (DNA and RNA) and the molecules that operate functionally to catalyze metabolic processes (protein enzymes). But we have recently come to understand that one of these three molecules, ribonucleic acid, in addition to being an information carrying molecule, like the DNA, shares with protein the ability to catalyze very specific chemical reactions (Kruger et al., 1982; Guerrier-Takada et al., 1983; Cech, 1986a; Zaug and Cech, 1986). Further, so far as we know, only ribonucleic acid has the ability to participate in both of these major events. Another way of putting this is that the central dogma distinguishes between genotype, the information coded, and phenotype, the expression of that information in the actual structures and metabolism of the cellular organism. But in RNA, genotype and phenotype are combined into one macromolecule (Woese, 1967). So we have partly revised the central dogma, and this has some important consequences.

An ordinary protein molecule, for example, an enzyme, is a polypeptide composed of a long sequence of amino acids. This chain folds up into a particular shape, and, as a result of that shape (what we call the conformation of the molecule), it can bind, for example, a small sugar molecule, glucose. By encouraging that molecule to undergo a specific chemical reaction, the protein serves as a catalyst or enzyme.

By contrast, an informational molecule, as we normally think of one, is the double helix of the deoxyribonucleic acid. It is a very long, thin molecule of repetitive structure. While this molecule seems ideally suited for storage of many bits of information, like a videotape, it seems poorly suited to wrap up into a complex shape, to attract small molecules and encourage the catalysis of reactions like a protein enzyme.

We seem to have the information on one kind of molecule and the catalysis and metabolism on a quite different kind of molecule. However, when we move from DNA to its cousin RNA, though it is chemically only slightly different, we have a different picture. Traditionally, there are three kinds of RNA: messenger RNA (mRNA), which serves as a template for protein synthesis; ribosomal RNA (rRNA), which is a part of the ribosomes on which the protein is assembled; and transfer RNA (tRNA), which is a small molecule that helps to attach the amino acids to the growing polypeptide chain in the order specified by the mRNA.

Let us look at transfer RNA more closely; it is one of nature's simplest RNA molecules. Transfer RNA functions as an interpreter between the nucleic acid language and the protein language. Although it still has double helical regions in its structure, the molecule folds to form a more complex shape. There are even regions here and there that form holes and pockets that look like they might be able to bind other molecules and catalyze reactions between them.

While transfer RNA in isolation has no known catalytic activity, other folded RNA molecules have been shown to be catalytic. These include the Group I self-splicing RNA discovered in my laboratory (Figure 1.1) as well as the RNA component of ribonuclease P, a tRNA-processing enzyme, which has been studied by Sidney Altman at Yale University and Norman R. Pace at the University of Indiana (Guerrier-Takada et al., 1983; Guerrier-Takada and Altman, 1984). Other catalytic RNAs such as the "hammerhead" have a catalytic core consisting of as few as a dozen nucleotides or so (Figure 1.2). In none of these cases has the complete three-dimensional structure that allows catalytic activity to take place been determined, but the two-dimensional pictures (Figures 1.1 and 1.2) already begin to portray the complexity of form that allows these molecules to catalyze rather sophisticated reactions.

One of the specific catalytic activities of the Group I RNA is the ability to assemble larger and larger RNA chains from small RNA chains (Zaug and Cech, 1986; Doudna and Szostak, 1989). This assembly process would never happen in a thousand lifetimes without the larger RNA molecule to facilitate the ligation or splicing of the chains, that is, to catalyze the reaction.

The detailed mechanism by which this catalytic RNA molecule rearranges chemical bonds to achieve an increase in chain length of the smaller RNA is in fact biochemically much analogous to the way that the contemporary enzyme RNA polymerase, which is a protein enzyme, cat-

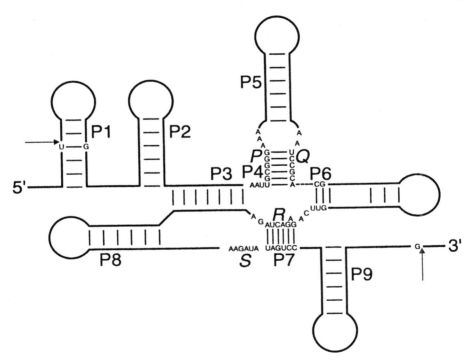

Figure 1.1 Conserved nucleotide sequences (*P, Q, R, S*) and base-paired structures (P1 through P9) of a naturally occurring catalytic RNA: The self-splicing Group I introns. The arrows indicate the sites at which RNA chain cleavage and rejoining are catalyzed (from Cech, 1988, by permission of Elsevier Science Publishers).

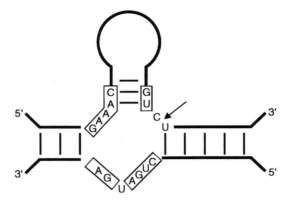

Figure 1.2 Conserved nucleotide sequences and base-paired structures of another naturally occurring catalytic RNA: The "hammerhead" ribozyme. The arrow indicates the site at which cleavage is catalyzed (from Cech, 1987, by permission of AAAS).

alyzes the elongation of RNA chains (Cech 1986b). This ability to facilitate the growth of other RNA chains is an activity directly related to questions of the origin of life to which I am coming.

How good a biocatalyst is one of these RNA molecules? Biochemists have several ways of assessing catalytic power. One calculation measures how fast the reaction occurs in the presence of the catalyst, relative to how fast it would occur spontaneously in an uncatalyzed reaction—if one just waited for the chemical bonds of RNA to rearrange in solution. They differ by some ten-billionfold (Cech, 1987). That represents the difference between a reaction that takes place in a minute, a time scale that is of biological relevance, as opposed to 19,000 years, which would be of more geological relevance. As a result of the catalytic action of RNA, we have a time scale that is compatible with life.

2. *The Origin of Reproduction*

So this brings us to life in the primordial soup (Oparin, 1957), or the primordial atmosphere (Woese, 1979), or perhaps in an organic scum at high temperature (Pace, 1991). We have to ask how all this got going, once upon a time. Can we give a more plausible picture now than before, now that we know that a single molecule can both carry information and accomplish the replication of new molecules that carry information?

If we are going to talk about the origin of life we are going to have to mix in some biological with some philosophical questions. "What is life?" There is no doubt that we have identified life when we find a fossil of an organism with a backbone and a digestive system, but how simple can life get and still be life? I am asking about the very beginnings. Can we have precellular life, or is that some kind of oxymoron? If there was once replication of information and catalysis, without cells, would that be life? I doubt that we can really talk about life until there came to exist a being with an inside and an outside, until something had separated itself from its environment.

There had to be some kind of a cell with a defining envelope. This envelope or membrane has tremendous biochemical implications. With it in place, the thing (we can now call it an organism) can take in nutrients and energy from the environment and sequester them for its own uses, including its own reproduction. We can begin to talk, in a sensible way, about natural selection proceeding when some organisms survive and some do not. In a way, the question is when do you have a self, some identity to conserve? When you have a wall around you? Or when you have information to replicate? That is more of a philosophical than a biological question; in biology both processes come together.

I am not going to give an account of how all the ingredients and processes of life, such as the cell envelope and the passage of materials across it, got started. We would need this for a complete definition of life. But still there had to be the precursors of life, and they had to be replicated. Maybe there is even a sense in which there is already life when nucleic acid molecules start replicating. I am only going to focus on replication and catalysis. These are absolutely essential to life and would appear in anybody's definition. There must be the ability to grow and to reproduce—the ability to generate more copies of oneself, to give rise to another generation of beings that have some similarity to the previous generation.

Since there is going to be, later in this volume, a good deal of worry about whether biological organisms have "selfish genes," perhaps I can say that, here at the start at least, this ability to replicate information does not seem to involve anything that could be called "selfishness" or that ought to be worried about ethically. Perhaps this evolves later in the higher organisms, such as the social animals, and it can be present in humans. But here at the early molecular level, all that is going on is the replication of information, and that is essential to any form of life. Perhaps there is here, almost at the start, something like a biological self to be protected, vital information to be conserved, but that process is part of the definition of life.

On the molecular level, reproduction means the replication of the informational molecules; in other words, copying the nucleic acid. Nature uses that system to assure that the next generation will look something like the previous generation, that it will have some of the same characteristics. So when we talk about the origin of life, we can focus, at least for my purposes here, on the question of the origin of replication of information. If we can solve that problem, then we have solved a major question about the origin of life. Again, I hasten to add, we will not have certainly solved the entire problem, but we will have solved the most fundamental problem about how life originates.

When scientists have thought about the origin of replication, they have looked at contemporary organisms for clues. What they see in modern living things is that nucleic acids cannot copy themselves, but must be copied by protein enzymes. At the origin of life, these same two functions needed to be fulfilled. But that would seem to call for two molecules, an informational molecule and a functional molecule to replicate the informational molecule. The chicken-and-egg problem of early evolution was, "What came first, the information or the function, the nucleic acid or the protein?"

This led to the view that perhaps, at the beginning, there must have been a very lucky juxtaposition, an extremely unlikely event, so that just the right nucleic acid molecule would be present at the same place and at the same time as a protein molecule, a molecule that would happen to be able to copy the informational molecule.

But now the picture is different. We know that one of the nucleic acids, ribonucleic acid, can play both roles—can have both the information-carrying capacity and the ability to catalyze chemical reactions. In fact, one of the RNA catalysts that we know about even has exactly the sort of activity that would have been required at the origin of life, that is, the ability to assemble other RNA molecules in an orderly manner, as described earlier. This leads to a considerable simplification of possible modes of origin of life. We can now postulate that, at the beginning, there was RNA serving both of these functions, and that other macromolecules like DNA and proteins were incorporated into the scheme at a later point.

Is such an RNA world plausible? There are two ways to test the theory. First, one can try to reproduce in the laboratory the individual steps that once, anciently, might have been the pathway to the origin of life and see if they are chemically reasonable. Do they occur rapidly enough and with high enough fidelity under reasonable conditions, such that it is plausible to think of life having begun in that way? After considering that test, I will turn to a second type of test, which involves so-called molecular fossils. One can ask, "Are there clues in today's organisms that point to life having begun with ribonucleic acid?"

So, first, is the self-assembly of RNA chemically plausible? To make an RNA molecule, one needs its building blocks, the nucleotides. Considering the atmosphere of the primitive Earth to have contained water vapor, carbon dioxide, carbon monoxide, methane, and ammonia and/or nitrogen, then in the presence of an energy source, such as sunlight or lightning, a number of small molecules like hydrogen cyanide, HCN, are formed. These then undergo spontaneous reaction with other HCN molecules, again in the presence of some radiation, like sunlight. We can simulate all of this in the laboratory, and, remarkably, two of the major products of these very simple reactions are two of the four nucleic acid constituents that we see in all living organisms today. These two are the adenine and guanine bases (often abbreviated A and G). The other two common bases, cytosine (C) and uracil (U), can also be formed in such experiments, although with more difficulty (Orgel and Lohrmann, 1974). Also, more than half of the biotic amino acids are formed in these prebiotic simulation experiments, so the building blocks for protein as well as RNA were around from very early on.

The next step might be to assemble the nucleosides, in which each base is attached to a sugar molecule. For RNA, the adenine or guanine has to be attached to a particular sugar molecule called ribose. For many years, the formation of ribose appeared problematic. When you start out with a simple molecule like formaldehyde and cook it up in a simple chemical reaction, you find that approximately one hundred different sugars are formed in various proportions, with no preference for ribose. Ribose is just one or two percent of the total mixture of sugars formed. But we need to use just one sugar, in this case ribose, if the assembly of RNA-like molecules is to take place. How could any system, in a relatively low state of order, sort one sugar out of all of this diversity of different sugars? If nucleic acid bases were joined to many different kinds of sugar, how could a functional nucleic acid ever be formed? It seemed rather that there would be chemical chaos (Orgel, 1986).

Often in these hypothetical origin-of-life scenarios, scientists come to a stopping point like this; and then, a couple of years later, someone discovers a solution. In this case, it was Albert Eschenmoser, an organic chemist at the Swiss Federal Institute of Technology in Zürich. He showed that if one simply changes the input materials, then, in a very simple chemical reaction, one can get a synthesis where more than half of the sugar is ribose (Müller et al., 1990). The success of such an experiment makes it appear that the intrinsic reactivity of these organic compounds is set up to give the nucleic acid that is needed for life, but we must be cautious: it is difficult to evaluate whether a reaction occurring in an organic chemistry laboratory is really under plausible prebiotic conditions. The reaction still produces two varieties of ribose (a racemic mixture of the two stereoisomers of ribose-2,4-diphosphate, that is, a mixture of two sugar compounds that are mirror images of each other) and also produces two varieties of phosphorylated hexose (six carbon sugar compounds) that form quite stable nucleic acids (Eschenmoser, 1991). So, from these experiments, a hexose nucleic acid-based origin of life seems at least as likely as an RNA world.

For polymerization into RNA, the nucleoside containing the base and the ribose sugar must join with an activated phosphate group. This also can occur under plausible prebiotic conditions, although once again there are unclear details (Orgel and Lohrmann, 1974). With the activated phosphate in place, then the nucleotides can be assembled into oligonucleotides, at first by random chemical processes and later by short scraps of nucleic acid serving as templates for the polymerization of complementary scraps (Orgel, 1986). So we have all the ingredients—the purines and pyrimidines, the ribose to make nucleosides, the phosphate to make nucleotides, polymerization to make oligonucleotides—and we are ready for a great moment in evolution, the origin of life.

How does catalytic RNA fit into all of this? There could be assembled, by random polymerization processes, a particular sequence that happened to have catalytic activity. This first, short string might have to be on the order of thirty individual monomer units in length, the size of the smallest RNA catalysts found to date, with perhaps a dozen of the positions requiring a specific nucleotide (Uhlenbeck, 1987). Interestingly, thirty is about the size limit of readily detectable products in Orgel's experiments. One can then envision self-replication getting underway (Figure 1.3).

The catalytic center, itself made of ribonucleic acid, would catalyze the joining of oligonucleotides. If one of the templates being copied was complementary to another RNA sequence that had replicating activity (a replicase), then a replication cycle could be established—extremely crude at first, but subject to improvement because of selective pressure. The dual nature of RNA—its ability to serve as both template and catalyst—is the key to self-replication (Sharp, 1985; Pace and Marsh, 1985; Cech, 1986b).

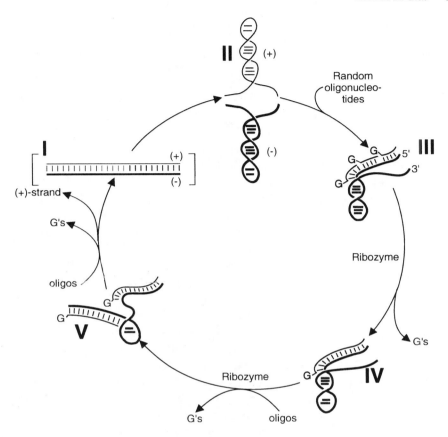

Figure 1.3 Scheme for RNA self-replication, the fundamental steps of which have been demonstrated under controlled conditions in the laboratory. Double-stranded RNA (I), consisting of a ribozyme (+)-strand and a complementary (–)-strand, undergoes denaturation-renaturation. The (+)-strand folds to form the active ribozyme (II). Oligonucleotides (oligos), random in sequence except for their 5′-terminal G residue, assemble on the template (–)-strand by Watson-Crick base pairing (III). They are ligated together by the ribozyme (IV). Ribozyme-catalyzed ligation continues (V), ultimately reproducing the starting RNA (I) and a new (+)-strand. Square brackets indicate that I is not an obligatory intermediate in the cycle (Pace and Marsh, 1985). In a subsequent round, one of the (+)-strands can serve as a template for the synthesis of another (–)-strand RNA. This cycle is derived from that proposed by Sharp (1985). (Reprinted from Cech, 1989, by permission of Macmillan Publishers.)

Biochemists are very close to achieving the reproduction of RNA molecules by spontaneous chemical processes in the test tube. Leading researchers in this area are Jack Szostak and Jennifer Doudna. They took a catalytic RNA molecule, one with highly evolved RNA replicase activity, and broke it into three pieces. The three pieces then self-assembled to give the active catalyst. That, in effect, starts the cycle. If the pieces could be replicated separately, they could then assemble into an active replicating unit. Then they could be dissociated again into individual pieces to serve as templates for the replication process. They tested one of the fragments for its ability to be replicated by the tripartite ribozyme, and attained reasonably efficient formation of a full-length copy over a period of about sixteen hours at 40 degrees centigrade, about the temperature of a warm bath (Doudna and Szostak, 1989; Doudna, Couture, and Szostak, 1991). Although this experimental system is very much simplified relative to any imaginable prebiotic environment, it serves to illustrate how the fundamental steps required for RNA self-replication are now rather well understood.

What are the advantages of RNA catalysis in this scenario? Why not just rely on template-directed polymerization and wait longer? First, because of their active site interactions, ribozymes can function efficiently with nanomolar, that is, extremely small, concentrations of RNA substrates (Herschlag and Cech, 1990) compared to the micromolar or higher concentrations required for simple template-directed polymerization. This is a real advantage when molecular building blocks are scarce. Second, the stereochemical peculiarities of a ribozyme active site can select chemically "correct" building blocks from a diverse population, ignoring the ones that might disrupt the regular structure of the polymeric product (Young and Cech, 1989), and string them together with exclusively the standard $5' \rightarrow 3'$ linkage. Finally, ribozymes greatly accelerate the rate of polymerization.

It might be thought that just by changing the speed at which assembly events take place, life still remains improbable. But the key factor here is the ratio of assembly to breakdown. The activated precursors to the nucleic acids are rather unstable, as are the bonds between the monomer units that make up the ribozyme polymer. If the rate of synthesis, or assembly, is not fast enough to keep up with the breakdown, then entropy will take over. The disorder of the system will win out. Molecules will come apart much more rapidly than they are constructed, and nothing will build up. But ribozymes step up the composition rate while the uncatalyzed decomposition rate is a constant, under any particular set of temperature and other environmental conditions. That makes the origin of life that much more likely.

3. *Molecular Fossils*

Such laboratory simulations are done, of course, with present-day molecules. Can we say whether today's RNA molecules resemble the original informational molecules? We do not have any fossils of the earliest such molecules. There are some physical fossils of early life, but these show us only the exterior surfaces. We do not know what went on inside those early cells. Nor do we have any physical fossils of precellular life, of molecules as they existed before they were incorporated into cells.

So is there any hope, given the lack of physical fossil evidence, of knowing the details of the origin of life? Are we left only to imaginative reconstruction experiments, simulations evaluating the likelihood of a particular origin-of-life scenario? Fortunately, there is another line of evidence. We might be able to reconstruct early evolution by scrutinizing modern cells for what are called "molecular fossils." Molecular fossils, according to Alan M. Weiner, a molecular biologist at Yale University, are structures or functions "that are shared by such diverse organisms that they must have been present in the first cells, as well as in earlier precellular forms of life" (Weiner, 1987).

Weiner's definition of a molecular fossil contains an argument that widespread distribution can only arise from common ancestry. The assumption is that these forms have been inherited over time in a branching, descending phylogenetic tree. However reasonable this assumption might seem, we cannot base a fully conclusive argument on widespread distribution because it is now well known that some genes have ways of hopping around. Genes can move with horizontal as well as vertical transfer of information, across and not just down the species lines. Because of jumping genes, widespread distribution falls short of being conclusive evidence that a molecule is primordial, present right there from the start. Nevertheless, if something is that widespread among all organisms on Earth, it seems plausible to think that it must have been present in the first cells as well as in precellular life. That all living organisms are derived from that early organism would then be a reasonable explanation of why these molecular fossils are so widespread.

There is another reason why the molecular fossil argument, though quite reasonable, is less than compelling: convergent evolution. Perhaps some biologically widespread molecule or process is the only, or easiest, or most probable solution to a biological problem. Independently, different organisms came upon this same solution. Paleontologists believe, for instance, that sight has evolved several times (Salvini-Plawen and Mayr,

1977); these often seem to represent wholly independent developments, not linked to one another in any way—anatomical, embryological, or evolutionary. Yet arthropods, mollusks, and vertebrates all have a related biochemistry of vision involving carotenoids (Wald, 1974). Similarly, today's RNA molecules could in principle be universal because, after life got started using other molecules, information transfer converged on the molecules that we now find so widespread.

Notwithstanding these uncertainties, let us look at one example of a molecular fossil that makes it plausible to claim that life originated with RNA. In all of the cells in our bodies and in all other contemporary organisms, many enzymatic reactions are facilitated by nucleotide cofactors, such as NAD^+, FAD, and co-enzyme A. These are small molecules used right at the core of a protein enzyme. Each of these contains a component clearly related to ribonucleic acid, an adenine nucleotide. This nucleic acid component is used by the protein enzyme as a handle to hang on to this reactive cofactor, which it uses to catalyze a biochemical reaction.

Why would the cofactors found in all contemporary living forms have this ribonucleic acid constituent? A reasonable answer seems to be that these are derived from the original RNA world, from a time when RNA was both informational molecule and biocatalyst. Then, as proteins started being synthesized in a template-dependent manner, there was a transition from the earliest RNA world to one where RNA and protein worked together. In that transition, these cofactors were retained. Later in evolutionary history, protein enzymes took over much of the role of biological catalysis. Today we can read the record backward, interpreting the co-enzyme as a molecular fossil of the active site of an ancient RNA enzyme. These cofactors are but one example of molecular fossils that can be used to argue the preeminence of RNA in prebiotic or early cellular life (Weiner, 1987).

4. *Life's Origin, Life's Value, and Human Responsibilities*

I will conclude with some thoughts about the philosophical or ethical consequences of this view that life might have originated with self-reproducing ribonucleic acid serving both informational and functional capacities. I am less comfortable drawing these conclusions, since I am primarily a chemist and a biologist, but I realize that there can sometimes be philosophical implications to biological discoveries, and that this is likely to be the case for discoveries that provide new information about how life might have originated. Also, although I am not a philosopher, one does not work on such problems for many years without, at least occasionally, wondering about the implications of one's work for a larger world view.

One conclusion from this view is that we see life as being the likely, perhaps even the inevitable, consequence of chemistry. That would mean that, at least under the circumstances present on early planet Earth, life is not a rare or an improbable event. If you have very common gases, such as were present on the primitive Earth, and suitable energy sources, such as sunlight and lightning, then nucleic acid molecules will arise by these chemical processes. Some of these will have catalytic activity, sluggish and extremely error-prone at first. In those cases where the catalytic activity allows the nucleic acid to replicate itself, there will be natural selection for sequences that are better replicators. This solves the major problem in getting life started, which is how to reproduce the informational molecule.

Melvin Calvin, in writing about the chemical evolution of life, said that he believed that the earliest information coding that made life possible "arose not by accident but because of the peculiar chemistries of the various bases and amino acids." He went on to remark that he had found "the beginnings of evidence that . . . there is a kind of selectivity intrinsic in the structures" (Calvin, 1975). But he could only make some guesses about the evolution of autocatalysis and of information transfer, and it was difficult at that time to see how the cycle—protein enzymes making informational molecules which in turn make proteins—might have started. I said earlier that often in these hypothetical, origin-of-life scenarios, scientists come to a stopping point, and then a few years later, a new discovery provides a plausible solution. The discovery that RNA can be an enzyme provides a powerful example of the selectivity intrinsic to the chemistry of the constituents of RNA, as described by Calvin, and takes us further than Calvin could then see by simplifying the problem of getting started. It provides more evidence that life is to be expected as a consequence of chemical principles.

This kind of view is fully compatible with Niles Eldredge's view of an evolutionary history in which there have been catastrophic extinctions on repeated occasions. In all of the later mass extinctions, even though they were so catastrophic that most species perished, there would have been some surviving species that respeciated, and so life rebounded in the way that Eldredge has described from the paleontological record. But if one of these major extinction events had occurred in the early history of life, all living things might have been snuffed out. We do have evidence that the early environment on Earth was subject to heavy meteor and asteroid bombardment, for instance. Temperature and climatic conditions were harsh. So it is quite possible that life started and was extinguished. But then we would expect life to originate again.

I am much less sure that each of these origins would have been chemically identical. There is not only one way in which life could have happened. Just as we see multiple branched pathways in higher evolution, there could have been multiple branched pathways in precellular molecular processes. We have in our natural history one solution that persisted. Were life to be found on another planet in another solar system more or less similar to ours, it might be very different morphologically, and different at the molecular level as well. On the other hand, one would be surprised if the types of molecules and molecular processes were so different as to be completely unrecognizable.

Given such a view, does this make life any less special, any less worthy of admiration? That question has to be asked first about early life. Does the fact that life arises more or less naturally from chemicals make such life of any less value? Does the fact that it might have originated more than once and might have been snuffed out in the past mean that life is less special? The answer we give will figure into our larger, philosophical world view, although we cannot have any ethical obligations toward life now vanished. But the same question has to be asked of the life that exists on planet Earth today, since all life is descended from the early life that, once upon a time, started this way. Is the biodiversity we inherit less worthy of our protection? Should we feel less concern about protecting individual species from extinction?

Someone might answer, "Well, yes, this rather degrades my view of life, because now I see life as such a common occurrence. If the origin of life, the creative act, is not a rare, single act but rather highly probable or inevitable, then creation can take place in perhaps many different environments over similar time spans." But should we say that life ceases to be an extraordinary thing and becomes an ordinary thing, and this cheapens it? It is true that we sometimes think that ordinary things are of less value, or that events that happen naturally or regularly are not so special. But that is not always true, and perhaps, with the origin of life, that an event happens naturally or more than once does not make it any less special. Such an event can still be quite remarkable and valuable. It would still be so if it were inevitable and happened repeatedly. *Catalysis* is the word I have been using, and it is the right word from a technical biological perspective; but what is also involved here is what a philosopher or a theologian might call creativity.

Consider an analogy. Which is more special: VanGogh's *Sunflowers* or posters of that artwork that are printed by the thousands? In one sense, people prize the original creative act much more than they prize multiple reproductions of such an act. The reproductions are of less value, because they are just copies of the first creation. It is true that we have been considering how replication got started, and that after its start, subsequently life is recreated by replication, when information is copied from one generation to the next. (We set aside for the moment the fact that the replication occurs with variation, where life creatively diversifies.) The mere replication of the original appears less special than the origin of the replicating process. The ten-thousandth organism is less remarkable than the first original one.

But that is not really the proper way to think of it. If we come to believe that life originated more than once, the more appropriate analogy is to think not of cheap reproductions but of original re-productions. It is as if a great work could be re-created time after time. Even though forces in nature sometimes extinguish life, there are other forces that recreate it. Seen that way, I feel much the other way about the origin of life, that this makes life even more special. If intrinsic to these small organic molecules is their propensity to self-assemble, leading to a series of events that cause life forms to originate, that is perhaps the highest form of creation that one could imagine.

The philosophers in this anthology, as noted at the beginning, are going to insist that we must have an account of how possibilities become actualities, and Langdon Gilkey and Charles Birch will protest, in the name of Alfred North Whitehead, that we cannot have possibilities floating in from nowhere (Gilkey, Chapter 7; Birch, Chapter 8). At least from the perspective of a biologist, I have given an account of how possibilities did, in times past, become actual. When this happened, life originated with impressive creativity, and it does not seem to me that possibilities floated in from nowhere; they were already present, intrinsic to the chemical materials.

The second part of the question about valuing life is not about the past but about the present. Given the view that creation might have been a frequent occurrence, does this give us any less responsibility toward life on Earth today? Here I would argue very strongly that the answer is, Not at all. If life originated in the manner that I have described, this does not mean that we are relieved from any responsibility. There are no revised ethical consequences. We value life today for what it now is. We do not want to commit what philosophers call the genetic fallacy, and argue that because the origins of something are different from what we supposed, the present responsibilities have changed. Life has what value it now has independent of its history, independent of what it might have been once a long time ago.

Biologists sometimes ask a question with a philosophical interest: "If you ran the tape of life over, would you come up with something completely different?" Some say that the origin of such forms of life as reptiles and humans was a chance phenomenon, and I have already said that I think my view is compatible with Eldredge's views about catastrophic extinctions. But I have been arguing also that there is a high probability or even an inevitability that nucleic acids will be formed and self-assemble and lead to life. I have said that life might have originated more than once; it is a kind of consequence of the chemistry. So is life an inevitable event? Or is it a chancy event? Or both? And how does this bear on its value?

Those of us who study biology are certainly convinced that different biological systems have found quite diverse solutions to similar problems. So it is true that we cannot imagine running the tape again and coming up with a natural history that is just like the one we have now. There is too much contingency in the natural selection. There probably were catastrophic extinctions in which many forms of life perished due to causes that had little to do with their natural selection. But there is a more fundamental question: Would there be some consistency at a very fundamental level—at the level of amino acids, nucleotides, nucleic acids, at the level of RNA? I am not certain about the answer, but I am of the opinion that there would be fundamental similarities in the basic life processes, which would then be covered over by tremendous numbers of somewhat more superficial differences.

There would have to be some autocatalytic process, assembling smaller units into larger ones so as both to store information and provide metabolic function. The importance of an envelope, a cell membrane, is so fundamental that, even though the composition of that membrane might be different, if we ran the tape again, there would be some sort of cellularization. The composition of the metabolic parts could be rather different and still be functional. And, of course, what I have sketched is an information storage process that permits many different kinds of life to evolve. Even if the information storage processes were similar, the kinds of species that evolved could be very different. Again, the fact that some basic similarities would recur, along with many differences, if it happened all over again, does not affect the value of the life that we now have on Earth. Both the differences and the similarities are part of what we value.

If life in the past originated in such a manner as I have described, this means that if we humans should somehow succeed in destroying life on Earth by accident or even deliberately, life would probably originate again. Indeed, the origination of new life forms might be going on right now, all the time, although the initial conditions are so different (for example, an oxidizing rather than a reducing or nonreducing atmosphere) that this cannot be assumed with confidence.

But the time frame is all wrong for this to affect our valuation of the life we now have. It presumably takes a long time to go from the first self-replicating molecules to the first life forms. It takes billions of years to advance to the point of having human life. So, as we destroy many species on Earth, and even in the extreme case that we might destroy all species including ourselves, yes, life will come back. But it will take millions or billions of years, a time frame that makes the reappearance of life of no particular consolation to us. The fact that life might come back in a billion years, and that it might come back differently, does not mean we have lessened responsibility today. Our responsibility has to be directed toward a time frame that is within our historical range—something that we can perceive, on the level of hundreds, thousands, or perhaps tens of thousands of years, not of millions or billions of years. Origins are one thing, and present responsibilities another.

Acknowledgment

I thank Holmes Rolston for his help in editing the transcript of my talk at the conference on "Biology, Ethics, and the Origins of Life" to generate this paper.

References

Calvin, Melvin. 1975. "Chemical evolution." *American Scientist* 63:169–177.

Cech, Thomas R. 1986a. "RNA as an enzyme." *Scientific American* 255:64–75.

Cech, Thomas R. 1986b. "A model for the RNA-catalyzed replication of RNA." *Proceedings of the National Academy of Sciences USA* 83:4360–4363.

Cech, Thomas R. 1987. "The chemistry of self-splicing RNA and RNA enzymes." *Science* 236:1532–1539.

Cech, Thomas R. 1988. "Conserved sequences and structures of Group I introns: Building an active site for RNA catalysis." *Gene* 73:259–271.

Cech, Thomas R. 1989. "Ribozyme self-replication?" *Nature* 339:507–508.

Doudna, Jennifer A., and Jack W. Szostak. 1989. "RNA-catalysed synthesis of complementary-strand RNA." *Nature* 339:519–522.

Doudna, Jennifer A.; Sandra Couture; and Jack W. Szostak. 1991. "A multisubunit ribozyme that is a catalyst of and template for complementary strand RNA synthesis." *Science* 251:1605–1608.

Eschenmoser, Albert. 1991. "Warum Pentose- und nicht Hexose-Nucleinsäuren?" *Nachrichten aus Chemie Technik und Laboratorium* 39:795–806.

Guerrier-Takada, Cecilia, and Sidney Altman. 1984. "Catalytic activity of an RNA molecule prepared by transcription *in vitro*." *Science* 223:285–286.

Guerrier-Takada, Cecilia; Kathleen Gardiner; Terry Marsh; Norman Pace; and Sidney Altman. 1983. "The RNA moiety of ribonuclease P is the catalytic subunit of the enzyme." *Cell* 35:849–857.

Herschlag, Daniel, and Thomas R. Cech. 1990. "Catalysis of RNA cleavage by the *Tetrahymena thermophila* ribozyme. 1. Kinetic description of the reaction of an RNA substrate complementary to the active site." *Biochemistry* 29:10159–10171.

Kruger, Kelly; Paula J. Grabowski; Arthur J. Zaug; Julie Sands; Daniel E. Gottschling; and Thomas R. Cech. 1982. "Self-splicing RNA: Autoexcision and autocyclization of the ribosomal RNA intervening sequence of *Tetrahymena*." *Cell* 31:147–157.

Müller, Daniel; Stefan Pitsch; Atsushi Kittaka; Ernst Wagner; Claude E. Wintner; and Albert Eschenmoser. 1990. "Aldomerisierung von Glycolaldehyd-phosphat zu racemischen Hexose-2,4,6-triphosphaten und (in Gegenwart von Formaldehyd) racemischen Pentose-2,4-diphosphaten: *rac*-Allose-2,4,6-triphosphat und *rac*-Ribose-2,4-diphosphat sind die Reaktionshauptprodukte." *Helvetica Chimica Acta* 73:1410–1468.

Oparin, A. I. 1957. *The Origin of Life on the Earth* (3rd ed.). Edinburgh and London: Oliver and Boyd.

Orgel, Leslie E. 1986. "RNA catalysis and the origins of life." *Journal of Theoretical Biology* 123:127–149.

Orgel, Leslie E., and Rolf Lohrmann. 1974. "Prebiotic chemistry and nucleic acid replication." *Accounts of Chemical Research* 7:368–377.

Pace, Norman R. 1991. "Origin of life—Facing up to the physical setting." *Cell* 65:531–533.

Pace, Norman R., and Terry L. Marsh. 1985. "RNA catalysis and the origin of life." *Origin of Life* 16:97–116.

Salvini-Plawen, L. V., and Ernst Mayr. 1977. "On the evolution of photoreceptors and eyes." In Max K. Hecht, William C. Steere, and Bruce Wallace, eds., *Evolutionary Biology* (vol. 10, pp. 207–263). New York: Plenum Press.

Sharp, Phillip A. 1985. "On the origin of RNA splicing and introns." *Cell* 42:397–400.

Uhlenbeck, Olke C. 1987. "A small catalytic oligoribonucleotide." *Nature* 328:596–600.

Wald, George. 1974. "Fitness in the universe: Choices and necessities." In J. Oró, S. L. Miller, C. Ponnamperuma, and R. S. Young, eds., *Cosmochemical Evolution and the Origins of Life* (vol. 1, pp. 7–27). Dordrecht, Holland: D. Reidel Publishing Co.

Weiner, Alan M. 1987. In James D. Watson et al., *Molecular Biology of the Gene* (vol. II, 4th edition).

Woese, Carl R. 1967. *The Genetic Code: The Molecular Basis for Genetic Expression*, New York: Harper & Row, p. 186.

Woese, Carl R. 1979. "A proposal concerning the origin of life on the planet Earth." *Journal of Molecular Evolution* 13:95–101.

Young, Benjamin, and Thomas R. Cech. 1989. "Specificity for 3′,5′-linked substrates in RNA-catalyzed RNA polymerization." *Journal of Molecular Evolution* 29:480–485.

Zaug, Arthur J., and Thomas R. Cech. 1986. "The intervening sequence RNA of *Tetrahymena* is an enzyme." *Science* 23:470–475.

2

Facing Nature

DORION SAGAN
AND
LYNN MARGULIS

■ ■ ■ *Editor's Introduction*

"Life," concluded Thomas Cech, "is a kind of consequence of the chemistry." Perhaps we should say the geochemistry. The elemental chemicals of life—carbon, oxygen, hydrogen, nitrogen—are common enough throughout the universe. But life is rare in the universe, though common on Earth, and the explanation lies in the special conditions in which these common chemicals find themselves arranged on Earth. Earth is ordinary elements in an extraordinary setting; there are super special circumstances that favor the events Thomas Cech has related.

Dorion Sagan and Lynn Margulis insist that, while nothing that Cech has said is wrong, all that he says, if it is to be correct, must be put in the context of the global Earth. Life comes indeed at the level of the Protista, the one-celled organisms that Margulis has made her specialty, but life is as intrinsic to the systemic Earth as to the microscopic materials. It is not just the microchemistry but the geochemistry in which the secret of life lies; the scale we must inquire about is not just molecular but equally planetary.

Earth is a lucky planet, with a set-up necessary for life, perhaps sufficient for the start-up of life. But for life to continue over the millennia, diversifying, ramifying, persisting, we have to keep our focus at the systemic level. Life remains as much a global phenomenon as it does a microscopic phenomenon. Sagan and Margulis, following James Lovelock, hold that Earth is a superorganismic system, named Gaia. True, the replicating molecules, once launched, continue at their microscopic levels. They become incorporated into cells, and these cellular systems, organisms, come to defend their "selves," which is simply to say that they defend their form of life. But we have also to reckon with how, once the life forces are underway, they to a marked extent remake the environment out of which they first came, molding it further in a prolife direction.

Life is not simply buffeted about by environmental vicissitudes; it is not passive before the geological and meteorological forces, but is interactive with them. The soil with its humus results from what otherwise would be only mineralized earth. The atmosphere with its oxygen, carbon dioxide, and ozone-shielding is a product of plant and animal life. Life modifies the climate. There are feedback loops set up between the organic and the nonliving world. The physicochemical environment is rebuilt biologically.

The phenomenon of life on Earth has to be seen as the bottom levels (the Protista) fusing to form upper levels, multicelled organisms, species, ecosystems, Gaia. This begins at the level of the mitochondria, essential in powering eucaryotic cells, and the plastids, essential in photosynthesis, which, again, power all life. These evolved from bacteria and algae that were long ago incorporated into cells. Thus eucaryotic cells (cells with nuclei) are each really populations, composed of vestiges of organisms that interact within the boundaries of the cell membrane. Higher organisms are communities, superorganisms themselves. Sagan and Margulis insist that there is no cause to cease thinking symbiotically at the level of (what biologists usually take as) individual organisms. Identity is a fluid thing: organisms are what they are within species populations; species are what they are in ecosystems, ecosystems are what they are in biomes, and so on until we culminate in the global Gaia. Symbiosis, no less than survival of the fittest, is the major driving force in evolution.

Sagan and Margulis dramatize this and connect with an ethics of respect for life by calling Earth the place with a "face," their way of getting at the difference Earth makes among the planets, asteroids, comets, and stars. Earth is place with life, a home more than it is just a place, a place with a presence. We now move from the molecules of Cech with their information replicated, from the Protista that Margulis also celebrates, to the global Gaia met and encountered. Sagan and Margulis see in the natural history of Earth a face, a presence, "an epiphany" (p. 46). Faces go with a kind of identity, with a self; but, interestingly now and in contrast to what (in later contributions to this volume) will appear to trouble ethics, this face does not produce a defense of self at all costs; it does not generate only selfish genes. Rather, there are fusions of identities until, at length, there arises a responsibility to Earth as a whole.

In the course of their claims, Margulis and Sagan will speculatively test various ideas—"facing" Earth, "forgetting" amoral natural history, accepting "superorganisms in history," "the Lucifer principle"—at risk of getting lost in the margins of philosophy, and with the opportunity of advancing the frontier. Biologists will want to remember, especially about Lynn Margulis, that she has been right before about symbiosis and synthesis when most of her colleagues thought she was wrong. But what are we to make of that phenomenon philosophically as we form a concept of nature? As we face opportunities for cooperation in ethics, transcending an essentially amoral nature?

Sagan and Margulis see collectives, superorganisms, as a kind of necessary, amoral evil (if also at the same time beyond good and evil) that uses individuals in the collective good ("the Lucifer principle"), "nature's 'builder,' an essential mechanism in evolutionary self-organization" (pp. 58–59). This phenomenon is similarly found in national and other social collectives that sacrifice individuals in their cause. Is this a phenomenon that we must endorse? That we ought to endorse? Ecologically, organisms can exist only within species and within ecosystems. Socially, humans can exist only within their heritages and social institutions. Perhaps this folding of individuals into larger historical lines can be cast in a different light? Perhaps Sagan and Margulis do this when they also find that "organisms work together in symbiotic truce for the mutual good" (p. 59) and think of this as the "ethical frontier"? If "partnerships conquer" (p. 59), then is seeing nature as nothing but selfishness and violent destruction of one form of life by another a "deviation from the symbiotic norm" (p. 59)? If "merging to form larger individuals is the way of the world," (p. 60), then perhaps we humans do have something to learn from amoral nature about the ways in which we can and ought behave? "Every self is collected into other selves" (p. 60). Is that a bad thing?

Sagan and Margulis recommend an active forgetting of our amoral origins in natural history. But what do we forget and what do we remember and respect about Gaia? Though we forget the amorality, they claim that we cannot forget the superorganisms that sacrifice individuals to the larger collective processes, and that we ought not to forget the symbiosis that is essential to creating more complex forms of life. Indeed, on the global scale, the cumulation of these syntheses into Gaia is something that we do not want to forget at all, but rather to face.

Is Earth the planet with a face? Or is that too misty an idea to be combined with the personal relationships, face to face, on which we normally attempt to build an ethics? Is the idea descriptive? Scientific?—as they want to claim about the Gaia hypothesis. Whatever one's conclusions, do we not need a model, a vocabulary by which we can reckon with life as a global phenomenon, by which we can respect Earth as a home planet? Nor is this just a rational need. Recalling the experiences of the astronauts on behalf of us all, Sagan and Margulis find that it is difficult to see Earth whole from space without an experience that combines biology and philosophy into ethics.

Edgar Mitchell, viewing Earthrise from the moon, wrote, "Suddenly from behind the rim of the moon, in long-slow motion moments of immense majesty, there emerges a sparkling blue and white jewel, a light,

Figure 2.1 The face of the Earth reflected in the visor of an Apollo astronaut. From NASA's poster "Apollo 11, 1969–1989." (National Aeronautics and Space Administration, Washington)

delicate sky-blue sphere laced with slowly swirling veils of white, rising gradually like a small pearl in a thick sea of black mystery. It takes more than a moment to fully realize this is Earth . . . home" (quoted in Kelly, 1988, at photographs 42–45).

Reference

Kelly, Kevin W. 1988. *The Home Planet.* Reading, MA: Addison-Wesley.

Dorion Sagan and Lynn Margulis have been prolific coauthors in biology and its interpretation. Recent works include *Microcosmos: Four Billion Years of Evolution from our Microbial Ancestors* (1986), *Mystery Dance: On the Evolution of Human Sexuality* (1991), and *What Is Life?* (forthcoming). Dorion Sagan is a writer and author of *Biospheres* (1990). Lynn Margulis is Distinguished University Professor of Biology at the University of Massachusetts at Amherst. Her books include *Origin of Eucaryotic Cells* (1970), *Symbiosis in Cell Evolution* (1981), and *Early Life* (1982). Her guest lectureships include the Twenty-Fourth International Geological Conference in Montreal (1972); the Royal Society, London (1978); Universidad de Barcelona (1982); and the American Association for the Advancement of Science (1991). Her work has been featured in profile stories in *Science*, *Smithsonian*, *Time*, and *Newsweek*. She is a member of the National Academy of Sciences.

Acknowledgment

The authors gratefully acknowledge support of this work by the Richard Lounsbery Foundation, New York; NASA Life Sciences, and the Dean of the College of Natural Sciences and Mathematics at the University of Massachusetts, Amherst.

But what is produced here is not a reasoning, but the epiphany that occurs as a face. (Levinas, 1969, p. 196)

But it seems that the demand for an ethics can only be satisfied by denying the ethical relation. It is as though the thinker were to respond by offering tablets of stone. It is, of course, no better a response to issue the instruction which refers ethics to the truth of Being. And yet it is at least the case that to refuse the demand is not necessarily to deny the relation. . . . In other words, the ethical relation occurs in the face-to-face relation, as witnessed in the demand for an ethics itself, a demand which it is as impossible to satisfy as it is to refuse. (Bernasconi, 1987, p. 135)

1. *An Ethics of Earth?*

Emmanuel Levinas has explored the ethical relation with the motif of "the face." From birth we respond to the face of our mother or caretaker. Infants copy the expressions, the smiling, the look of concern, the serene gaze, or tongue-wagging tomfoolery of adults. In this essay, we explore some possibilities opened by thinking the face in relation to evolutionary ethics. Is it possible to consider the biosphere a sensitive, responding subject? Would we be more responsible to the planetary environment if we saw the Earth as a being endowed with a face?

Thought of as a great transhuman subject, the living Earth, sometimes known as Gaia, becomes more than the "Earth" of Earth sciences or the "geo-" in geology. Its status as object is transcended. In a Gaian ethics, the biosphere as entity may provisionally be given the status of transhuman subject. Biodiversity, the atmosphere (with its chemically anomalous mixture of gases), boggy wetlands with abundant sphagnum and other moisture-retaining life, even apparent rocks which are in fact stromatolites made by bacteria—all these are more than mere "things": they describe beings or parts of living beings. Indeed, rather than simply physics and chemistry, the Gaian view of Earth uses physiology—geophysiology—as a describing science (Lovelock, 1988). Perceiving the Earth as a living being with a "face"—although not necessarily a human one—is one way to confer upon it ethical worth. It is one way to recognize the possible intersubjectivity of a planet made alive by its totality of beings.

Here we attempt no "prescriptive ethics" that would tell us what we should do. A prescriptive Gaian ethics, akin to planetary medicine or morality, must already know what a "healthy" or "good" human relationship with the rest of the biosphere is. After diagnosis of Earth's ills such a Gaian ethics would be prepared to prescribe solutions. The possibility for being responsible toward Earth seems heightened by treating it as an ethical being without personifying it. And this may be aided in some small measure by thinking of Earth in relation to the face. Moreover, despite initial appearances, the face of the Earth is not limited to that of the cartoon.

Analyzing Levinas' thought, which focuses not on the global-ecological but on the human-religious realms, Bernasconi finds a general problem with prescriptive ethics beyond particular maladaptive or unsatisfactory ethical prescriptions. A list of what you ought to do paradoxically absolves you from the responsibility of thinking ethically about the changing relation between and among beings. In a list of ethical "shoulds," the responsibility that occurs without prescription in the presence of the other's face is lost. "[T]he demand for an ethics can only be satisfied by denying the ethical relation" (Bernasconi, 1987.) Even writing, if it is to have the character of the ethical relation, cannot be entirely faceless. But in listing commandments on tablets of stone, the abstract law of a prescriptive ethics replaces the unsystematizable uniqueness of the face-to-face relation. "The ethical relation occurs in the face-to-face relation." For the face-to-face relation that grounds human interaction to apply to planetary life, Earth perhaps must receive some semblance of face. Conferring a face on an object putatively without one may seem to be a juvenile and unscientific personification. Yet even cartoons may produce or heighten an ethical relation: consider Spiegelman's Maus, a Pulitzer Prize-winning nonfictional account of the holocaust, rendered in comics for adults, in which Jews are portrayed as mice, Germans as cats, Poles as pigs, and Americans as dogs (Spiegelman, 1991, and see McCloud, 1993).

Levinas contrasts ethical "reasoning" with "the epiphany that occurs as a face." Faces come before and after words. There is a special power and work done by the face, according to the writer Rita Mae Brown: character is revealed in the visual as well as in the verbal medium; and the first appearance of a character's face ought to be dramatic, part of the characterization. When, in the late 1960s, the Apollo astronauts first viewed and photographed the face of Earth, there was an epiphany. Although repeated showing has made this face more familiar, the "face" of the Earth still is one that cannot be forgotten.

We seem to respond instinctively to animal faces. Slight variations in facial muscles signal the moods of apes and humans (Darwin, 1965). Humans are neotenous; adults of our species retain the characteristics of our ancestors as youths. People have "juvenile ape" traits such as large foreheads, small molars, and a love of play. We may have an inbuilt response to childlike faces even in other species. Stephen Jay Gould has traced the evolution of the animated character Mickey Mouse, concluding that cartoon watchers prefer the babylike modern Mickey to the less neotenous, more beady-eyed and ratlike depiction in earlier films. Cats and dogs, our favorite domestic pets, both possess distinct faces and relatively big heads and eyes. Indeed, nonhuman mammals respond with tenderness and care to warmth, softness, and youth expressed by newborn faces; dog mothers have been known to raise kittens, as have gorillas such as the famous gorilla Koko, who, taught American sign language, signed her desire for a live cat and, after receiving a minx named All Ball, came to love and care for it and call him her "baby" (Patterson and Cohn, 1985).

We dislike or distrust faceless creatures: fungi, jellyfish, even hooded criminals or burn victims. There is no responsibility without the promise of a face. Consider the modern techno-war, waged on the faceless from afar. Even the academic process of peer review where critics judge behind a cloak of anonymity, without name or face, encourages irresponsibility. Possession of a face is perhaps a prerequisite to responsible response.

Environmental science, coupled with a newer, Gaian discipline of geophysiology suggests Earth is a great body regulating its temperature, its oceanic salinity, its proportions of atmospheric gases and other variables (Margulis and Karlin, 1994). If the body of Earth is alive, but not human, nor even animal, then its face too will be neither simply human nor animal.

Levinas shies away from developing any ethical code of commandments because he conceives any such list to compromise the ethical relation. Ethics is based on the infinitude of the "face-to-face" relation with the Other, not on reasoning out an absolute code. Ethics is always open to what Levinas calls the moral summons of another. According to Levinas, "preexisting the plane of ontology is the ethical plane" (Levinas, 1969, p. 201). Jacques Derrida calls his an "Ethics of Ethics" (Derrida, 1978, p. 111). For Levinas, the face, whether in conversation or silence, is irreducible, a primordial ethical datum that responds and is responsible in real time. The face contrasts with the dead world of unresponsive facts:

> The work of language . . . consists in entering into relationship with a nudity disengaged from every form, but having meaning by itself. . . . Such a nudity is the face. The nakedness of the face is not what is presented to me because I disclose it, what would therefore be presented to me, to my powers, to my eyes, to my perceptions, in a light exterior to it. The face has turned to me—and this is its very nudity. It is by itself and not by reference to a system. (Levinas 1969, pp. 74–75)

We are born to learn faces; infants learn to recognize the face and have a propensity to perceive faces even in inanimate objects ("the man in the moon"). By undermining belief in God, Darwinian evolution upset traditional Western ethics based on Christian morality. Disturbing prescriptive ethics created a vacuum: What should humanity do? What sort of conduct is inherently or naturally legitimate in a world where the "fit"—those with the most offspring—survive? In a world where virtue is its own reward?

Aware that the basis of traditional prescriptive ethics had been removed, Herbert Spencer, a Darwin champion, drafted a "Synthetic Philosophy" whose aim was "that of finding for the principles of right and wrong, in conduct at large, a scientific basis" (Tillet, 1914, p. 4). With God set aside, Spencer, seeking the moral equivalent of a Euclidian geometry, yearned to ground ethics in science. For him, that meant evolutionary biology. He prefaces his ethics:

> Now that moral injunctions are losing the authority given by their supposed sacred origin, the secularization of morals is becoming imperative. . . . Most of those who reject the current creed appear to assume that the controlling agency furnished by it may safely be thrown aside, and the vacancy left unfilled by any other controlling agency. Meanwhile, those who defend the current creed allege that in the absence of the guidance it yields, no guidance can exist: divine commandments they think the only possible guides. . . . Both contemplate a vacuum, which the one wishes and the other fears. As the change which promises or threatens to bring about this state, desired or dreaded, is rapidly progressing, those who believe that the vacuum can be filled, and that it must be filled, are called on to do something in pursuance of their belief. (Spencer, 1890, p. viii)

Motivated by desire for safety, control, and guidance, Spencer writes of the terrors of an impending moral chaos opened by science, and the need to close it. But a prescriptive ethics based on the perceived modus operandi of evolution, survival of the fittest, leads to the abuses of child labor, social Darwinism, and Nazism—practices that have been legitimized, excused, or condoned as natural under the law of natural selection. Our own post-Nietzschean view is that "nature," unlike the God of the Old Testament, is beyond good and evil. Suspended in the moral vacuum, the discomfort we feel must be accepted as a spur, not so much to come up with a set of permanent ethical laws, but to respond with heightened sensitivity to individual face-to-face relations that elude generic prescriptions.

2. *Michael Ruse's Evolutionary Ethics*

Modern evolutionary ethics resists the fact that nature cannot replace God or the Bible as a ground of ethics, despite traditional morality's failure to provide a sturdy basis for its list of oughts in the wake of evolutionary theory. Consider, for example, a key figure in modern evolutionary ethics, Michael Ruse (see Chapter 4). Ruse distinguishes between the literal and an analogical use of biology in ethics (Ruse, 1988, pp. 71–78). An analogical use of biology to elucidate ethics sees evolution as progressive, analogously to progress in culture, and progressing toward, among other things, higher ethical and moral standards. Spencer held that more refined behaviors and ethical codes of civilized humans evolved from the general brutish conduct of beasts, the lower animals. He asserted that by examining this incomplete progress an abstract ideal of ethically responsible behavior can be formed. Ruse claims, instead, to make literal use of biology to elucidate ethics. For Ruse, morality is literally adaptive: "Just as we believe the Pythagorean theorem to be true, because it is in our biological interests to do so, so also we believe that we should love our neighbor as ourselves, because it is in our biological interests to do so" (Ruse, 1988, p. 74). "Our moral sense is an adaptation helping us in the struggle for existence and reproduction, no less than hands and eyes, teeth and feet" (Ruse, Chapter 4, p. 97). In Ruse's sociobiology-based view, the units of natural selection, genetic or individual, are fundamentally selfish. We help others only because they help us in return; there is no true selfless altruism but only enlightened self-interest. "Our moral sense . . . is a cost-effective way of getting us to cooperate" (Ruse, Chapter 4, p. 97). "The evolutionist's case is that ethics is a collective illusion of the human race, fashioned and maintained by natural selection in order to promote individual reproduction" (Ruse, Chapter 4, p. 101).

Our presumed innate disposition toward morality works more effectively because we and our own offspring survive, increasing fitness, if we deceptively overlook or forget about the amorality at morality's roots:

> In fact . . . biology often works better if we do not recognize the full story. . . . The essence of this approach to evolutionary ethics is that we will cooperate only if morality is genuine, however caused. If we were forever scheming about manipulating people to our own ends, under the guise of morality, we would be far less efficient cooperators than if we were (as indeed we are) genuinely moral. (Ruse, 1988, p. 78)

Ethics here is seen as a mask for selfishness. Yet Ruse's rhetoric is self-defeating, since the parenthetical avowal ". . . (as indeed we are) . . ." of our morality would redefine *genuine* as evolutionarily adaptive self-deception. Ruse seems to be holding that he himself (as part of the "we") is indeed moral, yet his theory demands that this morality not be exempt from evolution-fostered illusion, a contradiction that certainly raises the question of whether Ruse considers himself outside the adaptation-mandated illusion of morality he elucidates. Ruse here takes for granted, if only as momentary parapraxis,[1] our genuine selflessness and moral agency in a biological universe otherwise deemed to consist of amorally evolving selfish elements. It is as if the sociobiological view were so deterministic and ethically impoverished that it is in calling upon a higher morality based on free choice ("as indeed we are") that we can be rescued from the moral vacuum of an artificial morality produced by selfish genes! What does it mean to be "genuinely moral" if not to be free of the tendency to deceive ourselves as to our goodness because it adds to the longevity of our genes? Ruse argues against himself that no such genuine morality exists.

A close reading of Ruse reveals him open to Gilkey's critique (Gilkey, Chapter 7), namely that those who hold a sociobiologically based theory of ethics often themselves have, in fact, a high ethical view to which they are not entitled on the basis of their theory.

Ruse writes, in an essay entitled "Evolutionary Theory and Christian Ethics": "We are selfish brutes, it is true. But, laid on this is a genuine

---

[1]An action in which one's conscious intention is not fully carried out, owing to conflicting unconscious intentions.

sense of morality. We do good because we think it is good. The evolution-ist's case is that, precisely because we think the good is good, we function a lot better as cooperators, than if we were always looking for personal gain" (Ruse, 1989, p. 263). Paradoxically then, for an evolved sense of morality to function perfectly it must go so far as to convince even philosophers studying it that there is a "genuine" sense of morality, one which, smelling of Christianity, transcends even the artifices of genes that would deceive us into thinking we are good to promote the genetic inter-ests of the selfish individual. But when one confronts others as ends, not means, one confronts them, in the metaphor we have been developing, face to face in an ethical relation that, no doubt, has its moments of calcu-lated survival but which, in the end, cannot be whittled down to the nar-row interests of selfish genes. These genes, furthermore, are only theoretically abstracted from their coexistence within the individual and biosphere (Sagan, 1988). An Earth-gazing astronaut perceives the planet within an ethical (face-to-face?) framework that it would be both prema-ture and confusing to reduce to the level of a "real" sense of morality based on the deceptive machinations of selfish genes. Indeed, the ethical relation, the relation to and before the face of the other, would seem to precede the very distinction between truth and illusion which grounds Ruse's "genuine" morality based on tricky genes.

3. *Edward O. Wilson's Ethical Dilemma*

Edward O. Wilson has cast the problem of ethics as a double dilemma. "The first dilemma, in a word, is that we have no particular place to go. The species lacks any goal external to its own biological nature" (Wilson, 1978, p. 3). So Wilson repudiates any Spencerian ideal, any abstract of ethical perfection based on evolutionary progression. Humanity has no telos, moral or otherwise, and this disabuses us of the conceit that we hu-man animals have progressed to some higher point ("these last stages in the evolution of conduct are those displayed by the highest type of be-ing" [Spencer, 1890, p. 20]), from which moral heights we can look back and down on those animal ancestors from which we have evolved.

Indeed, although intelligent and populous, we, like all social ani-mals, have evolved means of interaction and mutual aid due to such pop-ulousness and intelligence. Fractally, our so-called "individual" bodies are comparable dense collectives of interacting life forms: eucaryotic cells. Our bodies display what might be called a totalitarian biology: all body cells die, "sacrificing themselves" for the whole, except the very few persistent fertilizing sex cells that, housing the genes, survive into the next generation. It is a deeply ethically troubling, and we feel not easily discarded, thought that the evolution of new, more inclusive individuals from crowds depends upon an increasing interdependence and dispensability of self-sufficient life forms. If there is such evolutionary method in the madness of crowds, it will not do to explain the worst excesses of this century away as simply a human aberration.

A second dilemma confronts us as we realize we can choose what to use, what to mitigate, what to discard from our pre-ethical biological heritage. Wilson continues:

> In order to search for a new morality based upon a more truthful definition of man, it is necessary to look inward, to dissect the machinery of the mind and to retrace its evolutionary history. But that effort, I predict, will uncover the second dilemma, which is the choice that must be made among the ethical premises inherent in man's biological nature. . . . Human emotional responses and the more general ethical practices based on them have been programmed to a substantial degree by natural selection over thousands of generations. The challenge to science is to measure the tightness of the constraints caused by the programming, to find their source in the brain, and to decode their significance through the reconstruction of the evolutionary history of the mind. . . . Success will generate the second dilemma. . . . Which of the censors and motivators should be obeyed and which ones might better be curtailed or sublimated? (Wilson, 1978, pp. 4–6)

But now Wilson thinks we can, after all, find somewhere to go. He has to break a way out of his dilemma:

> Because the guides of human nature must be examined with a complicated arrangement of mirrors, they are a deceptive subject, always the philosopher's deadfall. The only way *forward* is to study human nature as part of the natural sciences, in an attempt to integrate the natural sciences with the social sciences and humanities. . . . Only hard-won empirical knowledge of our biological nature will allow us to make optimum choices among the competing criteria of *progress*. (Wilson, 1978, pp. 6–7, emphasis added)

Wilson argues that a knowledge of our mammalian and hominid past is prerequisite to the formation of any scientifically grounded ethics. This past, seen first as a set of limitations, becomes to him a set of possibilities for moving forward ethically. When we know some of these aspects of our evolutionary past that limit us, we can confront Wilson's second dilemma and begin selecting from the menu of inbred behaviors. But, Wilson's hopeful call for science to ground ethics may never be answered. In the realm of ethics, science seems to take away as much as it gives. Consider two examples where science, that is, knowledge of a biological problem, might be expected to offer a definitive solution but in fact does not. A first example of a biologically limiting, ethically troublesome trait within the human phenotype is synthesis of the androgen testosterone. Testosterone's role in masculinization is well attested; its relation to male competitiveness and aggressiveness is less well established. Knowing the implications of this hormone for male violence, we might think that we can lessen the chances of war, for example, by choosing women to place in positions of political power. Testosterone is genetically programmed, and a choice to avoid testosterone "poisoning" in our political systems might represent a vindication of Wilsonian evolutionary ethics. But there is a Catch-22: groups ruled by women may have succumbed precisely because they lacked the requisite aggressiveness to compete with male-run collectives. The testosterone that poisons may simultaneously enhance evolutionary success. With or without modern science, we are caught between the hard place of a Judeo-Christian morality privileging the weak (perhaps, as we shall see, for the benefit of society), and the rock of the less hypocritical and more ruthless Machiavellian morality of warrior nobles—a pre-Christian ethos in which it is possible to find anticipatory hints of social Darwinism. Lowering testosterone for the sake of world peace, in other words, would be a politically correct movement in the direction of moral progress as traditionally construed. And yet, if instituted, it would also probably be not only demasculinizing but enfeebling of the human species.

A more complex example of Wilson's dilemma is illustrated by Pierre L. van den Berghe's analysis of aggression. Van den Berghe (1974) provisionally traces human aggression to hierarchy and territoriality, both functions of population pressures related to resource competition. He marshals evidence that we are the most hierarchical of mammal species, with the highest level of intraspecific aggressiveness, including a strong tendency for war-making. Our successful habitation of virtually all land areas on a resource-limited globe derives in large part from these human aggressive tendencies, which are a general feature of all cultures. He takes this to be evidence of a biogenic, rather than cultural, origin for aggressive behaviors.

Van den Berghe concludes:

> Cross-specifically, aggressive behavior serves the basic function of gaining access to or defending scarce resources for which other members of the same species compete. . . . Territoriality and hierarchy . . . regulate competition and thus aggression [but] they also require a great deal of aggression, both to maintain and to change. . . . The obvious need now is to develop our capacity for controlled biosocial change. This we can do only if we first realize that the problem exists. If we are to have any chance of success in controlling the nefarious effects of our acquisitive territoriality, our status-striving, our aggressiveness, and our run-away fertility, we must understand better the biological parameters of the social behavior of our species. The recognition that our behavior is in part biologically determined is by no means a counsel of despair. . . . Animal breeders have long known how rapidly controlled biological selection can modify a species. The spectre of eugenics conjures staggering ethical problems, of course; but sticking our social scientific heads into the biological sand . . . is hardly a solution either. (van de Berghe, 1974, pp. 784, 788)

Here is a more generic, more terrifying horn of Wilson's second ethical dilemma. Van den Berghe finds that the hierarchy and territoriality, resulting from long-standing human aggressiveness, may need to give way as we self-tame our species. He suggests a eugenic solution that raises "staggering ethical problems." Unless we decide to the contrary, the cycles of war, induced by aggression, will continually reduce the human populations, followed again by overcrowding that begets competition that foments aggression and war. The ethically distressing root problem is the Hobbesian war of "all versus all." If war, as the epitome of the (social) Darwinian mechanism, is to be stopped, then natural selection, which simultaneously breeds more dangerous warriors, should be curtailed. In a sense, Judeo-Christianity seeks to impede war in its prescriptive ethical emphasis on protection of the weak, empathy for the poor and crippled, and so on. The partially inherited passion for fitness-enhancing violence is categorized under the prevailing morality as evil. Two thousand years of Judeo-Christian ethics cannot simply be replaced with the carte blanche of natural selection. The essential lack of ethical ground prescribed by nature and her ways is almost too frightening to face.

Nonetheless, Wilson remains an optimist, though his biological descriptions give him no cause for his optimism. His faith in social advance ("the way forward," "progress") protects against stark, irremediable ethical chaos. Can a prescriptive evolutionary ethics really rely on the "progress of science"? Apart from the uselessness of protesting technological breakthroughs, are not science, knowledge, and progress implicated not just in the use but in the structure of implements of modern warfare, paradigmatically nuclear weapons? Should not the ethical relation signified by the face be maintained as a privileged outpost in the increasingly faceless landscape of modernity, which includes the program to base a prescriptive ethics on rational plan and cosmic law?

4. *Facing and Forgetting Ethical Groundlessness*

We cling to the traditional idea that, confronted with behavioral options, a prescribed "right thing" in the direction of moral progress necessarily exists. Values associated with moral progress—empathy, courtesy, proscriptions against killing, stealing, adultery, and so on—may have evolved because they tend to strengthen social ties, augmenting the unity of amoral superorganisms (corporations, nation-states, religious cults) in which all of us to varying degrees are dispensable parts. Indeed, there may be no scientific ground on which to develop a universal moral code. This groundlessness resembles looking into the sun, so bright that one cannot stare at it for long.

Nonetheless, we must face, if we are to consider whether we ought to reject and forget, this ethical groundlessness that underlies us in our natural history. If organisms are indeed fundamentally selfish (or, what amounts to the same thing, if they are the collectively selfish products of those self-replicating nucleotide polymers that Thomas Cech describes in Chapter 1), then the evolution of life is describable as a power struggle among selfish entities. Winning is scored by producing descendants that persist in the next generation. If entities cooperate, this cooperation is also nonetheless a form of "ganging up," of becoming more powerful through increased numbers and the pooling of biochemical, energetic, and locomotive skills.

The resources needed for life on Earth are always limited. Nitrogen (for proteins), phosphorus (for ATP, DNA, and RNA), sunlight (for photosynthesizers), and mates (needed for sexual reproducers) are all so limited that life is struggle. Organisms feed on each other, capturing each other's resources. The vast biological reserves, the global commons of oceans and atmosphere, have themselves succumbed to human population pressures. Although extensive, such reserves are always limited relative to the tendency for exponential population growth.

One species of organism provides a source of food and energy for another; different kinds and members of one species compete aggressively with each other for finite matter. Predation and exploitation follow. The self-justification of even the most pernicious practitioners of social Darwinism is delivered within the context of life's Malthusian character, what Nietzsche called life's "will-to-power." "Every moment devours the preceding one, every birth is the death of innumerable beings; begetting, living, murdering, all is one" (Nietzsche, 1911, p. 8).

The biota—the sum of flora, fauna, and microbiota—as a whole biological being is self-predatory; biological being imperatively feeds upon itself. Tennyson recoiled from "nature, red in tooth-and-claw" (Tennyson, 1850), and such a nature greatly troubled Darwin (as Birch recalls, Chapter 8). When Milan Kundera (1988) contrasts the "horror of living" with the "horror of being," he is referring in the former case to the individual person, who must live toward a finite end (death), but in the latter to the more horrendous infinitude of being, presumably including life on Earth as a carnophagic, flesh-eating process so far without end. Despite the moral efforts of humans and probably other animals, the great flows of ecological being and evolutionary becoming are not ruled by moral considerations. These problematic interrelations between organisms trouble those tempted to invoke ethical absolutes or to search nature for an absolute sense of moral direction.

Nietzsche's depiction of life as fundamentally amoral, as "beyond good and evil," a depiction that Judeo-Christianity has thought false, resists the tendency to assume that a heirarchy of values inheres in nature prior to our human interpretation of it. The prophet Zarathustra, speaking for Nietzsche, urges us "to break the tablets." Nietzsche's biology-influenced ontology is frighteningly unhypocritical. In stark contrast to Spencer, who reeled from the ethical vacuum, Nietzsche risks the dangerous gaze into the blazing sun of ethical groundlessness.

Faced with the monstrosity of an infinite ethical abyss, perhaps the only solution is that of the "active forgetting of [biological] being." Derrida underlines this in his groundbreaking essay, "The Ends of Man": we are left contemplating the image of Nietzsche-Zarathustra climbing a mountain in solitude, laughing as he burns the very documents that record the dire realization of ethical groundlessness:

> We know how, at the end of Zarathustra, at the moment of the "sign," when *das Zeichen kommt*, Nietzsche, distinguishes . . . between the superior man (*höhere Mensch*) and the superman (*Übermensch*). The first is abandoned to his distress in a last movement of pity. The latter—who is not the last man—awakens and leaves, without turning back to what he leaves behind him. He burns his text and erases the traces of his steps. His laughter then will burst out, directed toward a return which no longer will have the form of the metaphysical repetition of humanism, nor, doubtless 'beyond' metaphysics, the form of a memorial or a guarding of the meaning of Being, the form of the house and of the truth of Being. He will dance, outside the house, the *aktive Vergesslichkeit*, the "active forgetting" and the cruel (*grausam*) feast. . . . (Derrida, 1968, pp. 135–136)

So perhaps we have to face and to forget this ethical groundlessness in our evolutionary natural history. But we still have Earth to face.

5. *Superorganisms and Evil*

What to do with this combination of nature beyond good and evil and our ethical responsibility, before the face of the other, to do the right thing? In *The Lucifer Principle: A Scientific Expedition into the Forces of History*, Howard Bloom fleshes out in popular idiom a biological perspective on history. He combines evolutionary epistemology and sociobiology with a critique of the theory of individual selection. Survival at the genetic, individual, and especially the group level compromises or relativizes both truth and goodness. History is driven by superorganisms that perform deeds that individuals construe as evil (such as war) but are necessary to the survival and/or expansion of larger entities. Bloom's examples of the violent behaviors of human collectives bound in supposedly moral fellowship include the cruelty of the Red Guard in the Chinese cultural revolution, the Islamic jihads, and the Christian crusades. History moves through biological time, and Bloom reinterprets the ethnocentric histories of nations as the ethically suspect histories of superorganisms:

The superorganism, ideas and the pecking order—these are the holy trinity of human evil. They are the primary trio behind human creativity and much of earthly good. But they are also the raw ingredients of The Lucifer Principle. . . . Ideas are the superorganism's backbone. Hatred produces the superorganism's skin—the barrier that separates it from the outside world. And humans roused to battle are the superorganism's teeth and claws . . . [ellipsis in original] . . . For each superorganism has its elements of good. But each is determined to see at least one opponent as the epitome of all that is bad. Each superorganism is determined to grow and gain glory . . . [ellipsis in original] at the expense of the social beasts it targets as its enemies. Nature has built these appalling directives into our very biology. . . .

But how does the Lucifer Principle relate to the world in which we live today? How does it affect our lives? Greatly. For we exist within the bowels of a superorganismic beast. America is a social creature—a cluster of individuals pulled together by a meme. . . . And it has underplayed the eagerness of other superorganismic beasts—listening to radically different memes—to snuff the American ideals out. . . . Our social creature was originally pulled together by ideas like liberty and equality. But it faces off against other superbeasts. These social brutes, too, use ideas as a backbone. And some of them hunger to rip the body of the organism in which we live limb from limb. (Bloom, 1989, pp. 96–99)

The Lucifer Principle is "the hidden pattern behind evil," and Lucifer, fallen angel, serves as the rhetorical motif around which Bloom interprets the superorganismic, biological basis of humanity's most vile tendencies.

The neo-Darwinian emphasis on individual selection does not undermine Bloom's analysis. The animal body itself is a collection of cells, a descendant of cellular superorganisms. Even sociobiologists have some room for superorganisms. Edward O. Wilson interprets social insects as superorganisms; Richard Dawkins in *The Extended Phenotype* (despite his insistence on genetic selfishness in his earlier book, *The Selfish Gene*), demonstrates global interdependence of spatially distant genes. And Elliott Sober, a distinguished philosopher of biology and contributor to this anthology, writes, "Groups and communities can be organisms in the same sense that individuals are. Furthermore, superorganisms are more than just a theoretical possibility and actually exist in nature" (Wilson and Sober, 1989).

With the discovery of superorganisms that direct history, Bloom recasts evil in biological terms as nature's "builder," an essential mechanism in evolutionary self-organization. Breaking away from Nietzsche, who abandoned the Judeo-Christian and Zoroastrian duality of good and evil, he relativizes evil. Bloom faces the problem of devising a biologically based, universalist, prescriptive ethics. If he does not stare into the sun of ethical groundlessness for so long as Nietzsche, neither is he ethically blind. Perhaps not only in history, but on Earth itself, do "we exist within the bowels of a superorganismic beast." Gaia, which includes us, survived for billions of years without us. The biosphere will continue after our speciation or demise.

6. *The Ethical Frontier*

Violence seems intrinsic to exponentially growing organisms in a global environment of limited resources. Even the most shining examples of symbiosis in evolution—the green parts, called chloroplasts, of plants and algae, and the oxygen-using parts in all plant, animal, and fungal cells— owe their genesis to conflict. Chloroplasts evolved when other cells devoured but did not digest them. Mitochondria may have come from invasive hungry *Bdellovibrio*-like bacteria that were permanently impeded in their attempts to destroy their hosts. Organisms work together in symbiotic truce for the mutual good. This superorganism strategy at times outstrips the simple destruction of those other beings that require common resources.

The neo-Darwinian, sociobiological perspective on evolution describes life as an irreducible struggle among selfish individuals. But life is more appropriately described as a cumulative series of sensuous interactions that include symbiotic adventures in which partnerships conquer. Nature alone is neither pure, moral, peaceful, nor selfish. Violence, disease, and predation have received from science disproportionate attention precisely because they are life-threatening deviations from the symbiotic norm. Yet no possible moral law can be conceived to eliminate the wild element in which organisms evolve new means of expropriating limited goods, including means dependent upon exploitation and destruction of other feeling beings. At best, cultivation of a sense of morality might curb such outbreaks.

Early evolution of microbial life was marked not only by destructive interactions but by symbiosis-mediated emergence of genetically distinct "individuals" residing within each other that formed new kinds of more complex "selves" with unique identities. One "self" comes to merge with another "self" and a new "self" arises. Biological identity is not fixed. Identities that flow into each other require constant reaffirmation. Individuals join to produce new identities at more inclusive levels. This kind of merging to form larger individuals is the way of the world, and applies also to humans (Stock, 1993). Over and over again in the history of life on the planet, larger beings that metabolically reassured themselves displayed distinct characteristics as they formed from smaller beings.

All organisms dwell in ecosystems. Embeddedness as physically connected symbioses occurs at ever larger levels. James E. Lovelock has interpreted all living things on Earth, humans included, as a giant geophysiology, a superorganismic system whose energy source is photosynthesis from the sun. The entire biota has a biological identity; symbiosis accrues at the planetary level to produce selfhood. Gaia is a planetary self. Unlike "individual" organisms, Gaia completely recycles its own wastes. Materially self-contained, she requires only the influx of solar energy. But this solar radiation, converted into the organisms that form Gaia's working parts—"her" thermoregulating, chemically self-adjusting, sensitive body—increases resources, wealth, food as organisms themselves, thus exacerbating the amorality of a regime without intrinsic rule or border.

These multiple levels and claims for selfhood muddy the waters when we try to describe ethics. Every self is collected into other selves. The imaginary boundary limiting the ethical collective always moves. In-group and out-group boundaries shift. Do unto others as you would have them do unto you. Yes, but then there are always others, outside the circle, who are not quite other: the slave, for example, who is not considered part of "We, the people," excluded from the alleged protection of the ethical precept. And when the excluded "others" press their case and sometimes overcome their former exclusion, as in the case of American slaves and women achieving suffrage, the ethical boundaries swell to encompass new members. Then the notion of the in-group is reformed again; immigrants arrive who are excluded from the community by the very groups who were once excluded themselves.

The borders of ethical responsibility transcend *Homo sapiens*. Is a fetus human? To what extent has it human rights? We destroy and exploit bovine mammals and gallinaceous birds beyond any universal human rights. Few advocates of animal rights include the snakes, cockroaches, mosquitoes, spiders, and bacteria within the fold of the ethically protected. Perhaps this is because they lack recognizable faces; they are easier to put in the "out-group." The line demarcating inclusion within an ethically protected community widens with time. The ethical frontiers bleed, expanding to the global community of life, the biosphere. Sooner or later, we face the Earth.

Fixing the ethical frontier—stopping the "bleeding"—is an ongoing challenge, one without end because it must encompass the vast finitude of living beings. Periodic enlargements reforming this ethical line will be made indefinitely into the evolutionary future, perhaps never doing justice to an ideal prescriptive ethics. The difficulty of arriving at a prescriptive ethics means that it is all too tempting to surrender the quest when, as we have tried to show, precisely this search lies at the heart of ethics, evolutionary or otherwise. Let us not reject the demand simply because the task is perennial. Breaking the tablets confronts us with the nudity of the face— or the "face," the engraved face, the gravid face of the mirror—scarred, dismembered, absolutely or mosaically. We see a planet. We see the face of a person fragmented in turbid, turbulent waters, a face seen whole once the ripples die out and the waters reflect serenely. Ethics may be only an illusion masking our drive for survival. Face or mask—that is still in question.

References

Bernasconi, Robert. 1987. "Deconstruction and the possibility of ethics." In John Sallis, ed., *Deconstruction and Philosophy: The Texts of Jacques Derrida* (pp. 122–139). Chicago: University of Chicago Press.

Bloom, Howard. 1989. *The Lucifer Principle: A Scientific Expedition into the Forces of History* (Book 5). Unpublished.

Darwin, Charles. 1965. *The Expression of the Emotions in Man and Animals.* Chicago: University of Chicago Press.

Derrida, Jacques. 1968. "The ends of man." In A. Bass, trans., *Margins of Philosophy* (pp. 109–136). Chicago: University of Chicago Press.

Derrida, Jacques. 1978. "Violence and metaphysics." In A. Bass, trans., *Writing and Difference* (pp. 79–153). Chicago: University of Chicago Press.

Kundera, Milan. 1988. *The Art of the Novel.* New York: Grove Press.

Levinas, Emmanuel. 1969. *Totality and Infinity: An Essay on Exteriority.* Pittsburgh: Duquesne University Press.

Lovelock, James E. 1988. *Ages of Gaia: A Biography of Our Living Earth.* New York: W. W. Norton.

Nietzsche, Friedrich W. 1911. *Early Greek Philosophy and Other Essays.* London and Edinburgh: T. N. Foulis.

Margulis, L. and E. Karlin. 1994. "Gaia, a new look at the Earth's surface." In E. Karlin, ed., *Ecology and the Environmental Crisis.* New York: HarperCollins.

McCloud, Scott. 1993. *Understanding Comics: The Invisible Art*. Northampton: Kitchen Sink Press.

Patterson, Francine, and Ronald H. Cohn. 1985. *Koko's Kitten*. New York: Scholastic Inc.

Ruse, Michael. 1988. *Philosophy of Biology Today*. Albany: State University of New York Press.

Ruse, Michael. 1989. *The Darwinian Paradigm: Essays on its History, Philosophy and Religious Implications*. New York: Routledge.

Sagan, Dorion. 1988. "What Narcissus saw: The oceanic "I"/eye." In J. Brockman, ed., The *Reality Club 1* (pp. 193–214). New York: Lynx Books.

Spencer, Herbert. 1890. *The Data of Ethics*. London: Williams and Norgate.

Spiegelman, Art. 1991. *Maus: A Survivor's Tale, And Here My Troubles Began* (Vol. II). New York: Pantheon Books.

Stock, Gregory. 1993. *Metaman: Humans, Machines, and the Birth of a Global Superorganism*. London: Bantam Press.

Tennyson, Alfred. 1850. *In Memoriam*, Part LVI, Stanza 4.

Tillet, A. W. 1914. *Spencer's Synthetic Philosophy: What it is All About. An Introduction to Justice, "The Most Important Part."* London: P. S. King and Son.

van den Berghe, Pierre L. 1974. "Bringing beasts back in: Toward a biosocial theory of aggression." *American Sociological Review* 39:777–788.

Wilson, David Sloan, and Elliott Sober. 1989. "Reviving the superorganism." *Journal of Theoretical Biology* 136:337–356.

Wilson, Edward O. 1978. *On Human Nature*. Cambridge, MA: Harvard University Press.

Mass Extinction and Human Responsibility

NILES ELDREDGE

3

■ ■ ■ *Editor's Introduction*

The origin of life is no single event, although initially there is the start up of replicating, cellular life, the primordial chemical evolution of information-bearing, metabolizing molecules. That launches life, but the journey lies ahead. Though in this anthology we are much concerned with that critical first chapter and with the critical latest chapter, the originating of humans with their ethical life, the natural history in between is quite relevant. The originating of life continues to take place over the millennia of evolutionary history. That is, indeed, most of the story, chapter after chapter of speciation and respeciation. Niles Eldredge, one of the world's leading paleontologists, next recounts "the extremely checkered career that life has had on Earth" (p. 70).

Eldredge is a Darwinian with misgivings. After the launching of life, natural selection begins, and—according to orthodox Darwinism—life evolves thereafter incrementally, with natural selection operating on random variations. That is the gradualism of natural selection, against catastrophism. Eldredge nowhere denies that natural selection is among the determinants of evolutionary history, but he is less sure of its centrality. Even in normal epochs, the usual result for a species is extinction, not a steady evolution into something else. "The search for a general ecological theory of extinction must refocus its gaze on earthbound causes" (p. 75). Climates regularly change at a pace with which life cannot cope. Sometimes, although habitats remain elsewhere on Earth, they are unreachable; and organisms have no possibility or genetic inclination to seek them out. "Failure to keep up, or simply to locate, such recognizable, accustomed habitat results in extinction, not evolution, as the overwhelming rule" (p. 77). When these climatic changes occur at a still faster rate, there can be catastrophic extinctions. The pace of natural selection is too slow to cope. The planetary geological forces are as indifferent, even hostile, to life as the astronomical ones. So Darwinian gradualism is deemphasized and the Earth story is recounted with more catastrophism. There have been "worldwide, nearly complete biotic turnovers," and Eldredge finds remarkable this "interplay between extinction and evolution that has so markedly colored the history of life" (p. 72).

But mass extinction is never the end of the story—at least not so far. Extinction is not the overwhelming rule after all; or perhaps we should say that life is not overwhelmed, despite the mass extinctions. Reflecting over the fossil evidence, epoch after epoch, Eldredge continues, "It has seemed to a number of biologists (particularly we paleobiologists) that Earth's biota is tough, able to rebound in both an evolutionary and ecological sense after even the worst of biotic devastations" (p. 68). Indeed, extinction is always coupled with respeciation, and this becomes a key to Earth's most novel epochs of respeciation. So evolution is the overwhelming rule, after all, except that it is not a gradualism. There is rather crisis and rejuvenation. The pace is what Eldredge calls "punctuated equilibrium" (Gould and Eldredge, 1977). Even normal times, to say nothing of catastrophic times, are punctuated by relatively abrupt mass extinctions with dramatic resettings of evolutionary history.

The origin of human life, too, arises in crisis, but, with the arrival of *Homo erectus* on Earth, evolutionary history is surpassed, owing to the dramatic "difference in being human" (Ayala, Chapter 5). "Now we begin to detect something novel in Earth's history, a species that can occupy many different environments, because it can remake its environments, owing to the advent of culture" (p. 80). The cultural acceleration of possibilities reaches a pitch in current events; humans are a species that can remake, and unmake, the world.

Once again, the possibilities are coupled with crisis. Caused now by technological prowess, rather than, as with catastrophic extinctions, by climatic change, we face another mass extinction. "We are presently in the midst of a global extinction event—one that threatens to rival the greatest of the mass extinctions of the remote geological past" (p. 67). And humans are the ethical species, with a duty to the natural world. But Eldredge is cautious about deriving ethics from biology because he knows too well how theories in biology have shifted over time. Dissatisfied with philosophical efforts to derive what *ought* to be in ethics from what *is* in biology, he is reluctant to derive more than self-interest out of biology, although he is also reluctant to think that this can be the whole basis of ethics. Even self-interest for humans as a class should be enough for biological conservation, however, since, at the global scale, humans have entwined destinies with the natural world.

Eldredge invites us to ask about the contingent versus the predictable in natural history, and to puzzle over how the same factors that cause extinction simultaneously promote respeciation, with its novelty. He asks whether destructions of habitats by humans, especially through agriculture, does not in effect mimic the habitat destructions that once followed from radical climate change, with both producing large scale extinctions. "We tend not to see ourselves in and of nature, but rather as conquerors of nature. . . . But we have not really and truly escaped nature" (pp. 82–83). Are we in, of, outside of, over nature? A global species? Or what? Do we need to "shed the illusion of separateness" (p. 83)? Can or ought we to "thrive . . . at the expense of the rest of the biosphere" (p. 84)? Anticipating the disputes to come later in this volume, can we get clear about the relation between our biology and our ethics, including our environmental ethics and our duties concerning other species? "We as individuals were born into, and as a collective species evolved in, a natural world that we ought to feel compelled to restore and protect. But *why* ought we feel that way" (pp. 68–69)?

Reference

Gould, Stephen Jay, and Niles Eldredge. 1977. "Punctuated equilibria: The tempo and mode of evolution reconsidered." *Paleobiology* 3:115–151.

Niles Eldredge is Curator, Department of Invertebrates, at the American Museum of Natural History in New York City. He is the author of *The Unfinished Synthesis: Biological Hierarchies and Modern Evolutionary Thought* (1985), *Macroevolutionary Dynamics: Species, Niches, and Adaptive Peaks* (1989), *Time Frames: The Rethinking of Darwinian Evolution and the Theory of Punctuated Equilibrium* (1985), *The Miner's Canary: A Paleontologist Unravels the Mysteries of Extinction* (1991), (with Marjorie Grene) *Interactions* (1992), (with Ian Tattersal) *The Myths of Human Evolution* (1982), and (with Joel Cracraft) *Phylogenetic Patterns and the Evolutionary Process: Method and Theory in Comparative Biology* (1980), as well as over 150 papers in paleontology. He edited the *Natural History Reader in Evolution* (1987). He is past co-editor of *Systematic Zoology*.

■ ■ ■ ━━

We are presently in the midst of a global extinction event—one that threatens to rival the greatest of the mass extinctions of the remote geological past. Though doubts persist as to the severity of our current predicament (Mann, 1991), most ecologists and systematists agree that the annual loss of species is higher than would be expected in normal, "background" extinction, and that the rate in all likelihood is accelerating (Soulé, 1986; Western and Pearl, 1989; Wilson, 1988). Ehrlich and Wilson (1991) estimate that at least 4,000 species are lost annually in the tropical rainforests alone.

The fossil record of the past 570 million years tells us clearly that such mass extinctions are indeed real and have repeatedly punctuated the history of life. Raup (1979) estimates that, in the greatest extinction event yet recorded, fully as many as 96 percent of all the species on Earth may have become extinct in a relatively brief span at the end of the Permian Period of geological time.

Similar as the extinction events of today seem to be to those of the remote geological past, analyses of the *causes* of extinction typically emphasize themes so disparate that, at first glance, there would seem to be little hope of formulating a single, overarching causal theory of extinction to embrace all cases, including the major and minor past events, as well as those enveloping the biosphere today. The blame for the current upsurge of species extinction is generally laid on our own collective shoulders: it is through the activities of *Homo sapiens*, it is generally agreed, that the world's biota faces its current crisis. But *Homo sapiens* has been around for, at most, 150,000 years. Even if we were to include our extinct ancestral and collateral kin species in an extended sense of "human," our record goes back only some 4 or 5 million years to a joint ancestry with the great apes. In contrast, the great mass extinctions of the past occurred in the tens and hundreds of millions of years ago, obviously without a helping human hand. The best of a truly formidable list of theories of such ancient extinctions invoke climatic change, habitat disruption and, most recently, the much-publicized possibility of collision between Earth and an extraterrestrial body.

Humans and comets would seem to have little in common as explanatory agents of extinction events. Yet, I shall argue that there is a single, unified theory of mass extinction that does indeed link up the events of the past with the present biotic predicament (Eldredge, 1991). Such a general theory is necessary, I believe, to achieve perspective on the role humans have played—and continue to play—in extinction.

Such a perspective, of course, reflects one sense of the term *responsibility* in my title. It has seemed to a number of biologists (particularly we paleobiologists) that Earth's biota is tough, able to rebound in both an evolutionary and ecological sense after even the worst of biotic devastations (a notion that agrees well with aspects of the notion of Gaia as developed by Sagan and Margulis, Chapter 2). From this perspective, it does seem a bit egotistical for us, members of *Homo sapiens*, to believe that our effects on the biospheric system of Earth can be so powerful as to single-handedly bring about a mass extinction event on our very own. Yet, as many have concluded before me, hubris aside, this is precisely what we seem to be doing. We are very much responsible for the present-day biodiversity crisis: I will conclude that our behavior in general mimics the effects of nonhuman causal agents of mass biotic destruction of the remote past. Moreover, we, as a species, have been in the extinction business for tens of thousands of years; we have simply expanded our effects on the biosphere in recent years.

But the "responsibility" of my title transcends simple recognition of our causal complicity in the biodiversity crisis. Recently heightened concern over environmental issues betokens an increasing willingness to **take** responsibility and actively search for effective measures to ameliorate the negative effects we have been wreaking on the planet. Here we begin to ask what we "ought" to do, and in so doing, enter the realm of ethics.

Conservationists point to several categories of reasons why humans should both care, and want to do something, about the loss of biological diversity caused by our own hand. For example, Ehrlich and Wilson (1991) enumerate three such general reasons: (1) ethical and esthetic, (2) utilitarian, and (3) maintenance of overall healthy status of the world's ecosystems. All have their merits—and in the decidedly pragmatic view that positive action, whatever the motivation, is salutary, all are to be commended. Personally, I would decouple ethics from esthetics: the emotional response of an individual human to undespoiled nature is very different from a set of ethical precepts drawn from particular conceptualizations of the nature of the biosphere. Rolston (1988) presents a particularly lucid review of ethical issues pertaining to the environment.

And, certainly, I subscribe to the view that we as individuals were born into, and as a collective species evolved in, a natural world that we ought to feel compelled to restore and protect. But *why* ought we feel that way? If the history of science tells us anything at all, it is that our views of nature are, by the very nature of the scientific process, bound to change—often radically. The concept of Gaia (Lovelock, 1987; Sagan and Margulis, Chapter 2) is relatively new. To some, Gaia implies that Earth and its surficial biosphere is very like a single living organism—which to some would imply inherent rights, plus our own duty to work for the health of the total system, much as a cell (in a healthy body) would, by performing normal tasks, contribute to the ongoing maintenance of an organism's body. Sagan and Margulis, though they speak of facing the Earth and of our responsibilities to respect life, do not ascribe "rights" to Gaia. They do, however, conclude that Gaia is a tough old bird, able to take whatever punishment that one of its minor constituents hands out to it.

Ethics—at least when derived from biological precepts—are very much in the eye of the beholder. The dog-eat-dog values of social Darwinism rested on (and were basically consistent with) a particular, rather narrow reading of Darwinian principles. Ruse (Chapter 4), in contrast, derives a view of ethics that effectively eschews the "selfish man" model of traditional social Darwinism, seeing, instead, cooperative behavior as the essence of our adaptive evolutionary heritage. I am in essential agreement with Ruse's conclusions about the adaptive basis of human social behavior (Caporael, Dawes, Orbell, and van de Kragt, 1989).

But Ruse reaches his conclusions because later Darwinians, modifying his view, claim to have settled the paradox of altruism by recognizing that genes are actually being selfish when one organism cooperates with another—because of a pay-off in the spread of an organism's own genes thereby achieved, as is most clearly the case when cooperation is among close relatives sharing calculable degrees of identity of genetic information. Thus, a minor gloss on Darwinian evolutionary biology results in an ethical system 180 degrees different from the social Darwinism of old!

I cannot but wonder at the form ethical systems might take when derived from more disparate versions of evolutionary theory! There is no necessary relation between biology and ethics—and whatever ethical considerations are brought to bear on the biodiversity crisis will remain, at least to me, unconvincing if they are derived in lawlike fashion from some particular view of biological ontology and history, whether expressly ecological or evolutionary in nature.

I have little to say about the utilitarian aspects of the case for human concern on the biodiversity issue. Loss of genetic diversity assuredly means loss of potential resources. The leukemia-fighting properties of the Madagascar periwinkle symbolize the untapped potential of many species now threatened by the habitat alterations that are still accelerating. Yet habitat destruction serves not only the greed of capitalism, but it is traded for the far more modest, yet essential, life-sustaining, economic needs of local peoples. Herein lies perhaps the area of greatest ethical concern raised by countermeasures against the mounting modern mass extinction: balancing the needs of the world at large with local human needs in targeted areas—whether those areas be tracts of Amazonian rainforest or the stands of old growth forest still remaining in North America.

Esthetics are largely personal. Ethics based on one scientific system or another are suspicious. The utilitarian argument itself raises ethical dilemmas. Given those problems, the overall health of the biosphere, of Gaia, emerges (for me, at least) as the most compelling reason for countermeasures against the human-caused biodiversity debacle. I shall delay developing this notion further, however, pending a general review of mass extinctions, the changing role of humans in nature, and the documented impact that humans have had as the root cause of extinctions in nature.

1. *Mass Extinctions of the Geological Past*

The geological time scale affords the simplest and most direct demonstration of the reality of mass extinctions (Figure 3.1). The very existence of such a chart, with all its subdivisions, reflects the extremely checkered career that life has had on Earth. *Paleozoic* means "ancient life," *Mesozoic* "middle life," and *Cenozoic* stands for "modern life." Two of the greatest mass extinctions yet recorded separate the old from the middle, and the middle from modern life eras on that chart. The earlier of the two, at the Permo-Triassic boundary, was apparently *the* most devastating: this is the event where, according to paleontologist David M. Raup (1979), perhaps as many as 96 percent of all species on Earth at the time perished. The later extinction, at the Cretaceous-Tertiary (K-T) boundary, is easily the most famous, as that was the event that took away the dinosaurs.

The geological time scale was pieced together by geologists in the late eighteenth and early nineteenth centuries. These early geologists were engaged in mapping strata, realizing that sedimentary rocks are the hardened products of a regular process of sedimentation, where grains ly-

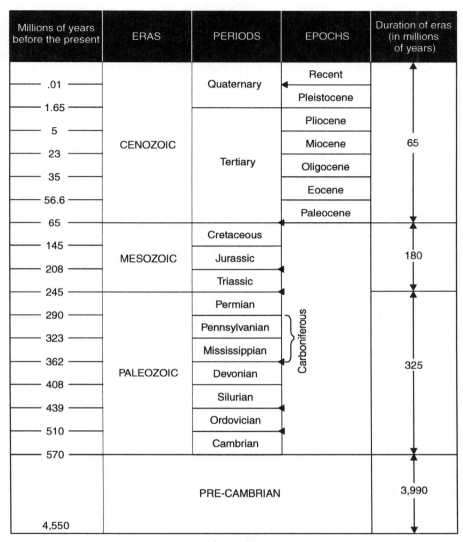

Millions of years before the present	ERAS	PERIODS	EPOCHS	Duration of eras (in millions of years)
.01	CENOZOIC	Quaternary	Recent	65
1.65			Pleistocene	
5		Tertiary	Pliocene	
23			Miocene	
35			Oligocene	
56.6			Eocene	
65			Paleocene	
145	MESOZOIC	Cretaceous		180
208		Jurassic		
245		Triassic		
290	PALEOZOIC	Permian		325
323		Pennsylvanian	Carboniferous	
362		Mississippian		
408		Devonian		
439		Silurian		
510		Ordovician		
570		Cambrian		
4,550		PRE-CAMBRIAN		3,990

Figure 3.1 The geological time scale. Black arrows indicate major extinction events. (From N. Eldredge. 1991. *Fossils: The Evolution and Extinction of Species*. New York: Harry N. Abrams, Inc.).

ing at the bottom are older than those lying above them. Because rock outcrops are discontinuous, these scientists looked for clues in the rocks themselves with which to recognize equivalent layers in different locales. Fossils became a favorite focus of study. Sedimentary rocks often appear to be "zoned," that is, divided into successive strata by characteristic suites of fossils—assemblages that always occur in the same relative order in a region. One of the pioneers, Baron Georges Cuvier (1812/1831), recognized many "revolutions" in the history of life. The faunas of each period were eventually erased by extinction, always to be replaced, according to Cuvier, by a separate creation.[1] Though most of the periods of geological time reflect the area in which they were first defined (Cambria, for example, is the Roman name for Wales), the terms *Paleozoic*, *Mesozoic*, and *Cenozoic* were in place by 1840—clear indication that early geologists were well aware that the history of life is, in effect, packaged into discrete segments with more or less distinct and characteristic life forms.

By the late eighteenth century, geologists realized that the grander subdivisions of the time scale could be recognized on the basis of fossils the world over; everywhere that paleontologists looked, they found the same basic scheme. The Permian Period, for example, is named for the Russian town of Perm in the western Ural Mountains. But it is also beautifully represented in the arid wastelands of western Texas, on islands in Indonesia, in China, the islands of southeastern Alaska, and so forth. And those divisions, to a remarkable degree, reflect turnovers of worldwide scope in Earth's ecosystems. The six most momentous of these ancient extinction events are indicated in Figure 3.1. All but one (in the Upper Devonian) coincides with the end of a recognized period of geological time; other divisions between periods (e.g. between the Silurian and Devonian) represent much less marked faunal turnovers.

Thus there is a spectrum ranging from worldwide, nearly complete biotic turnover down to minor episodes, local in extent and encompassing more modest percentages of species. What is remarkable is that more has not been made of these turnovers and consequently of the interplay between extinction and evolution that has so markedly colored the history of life. That we paleontologists are only now lavishing such attention on extinction is a reflection of two things: the development of the Alvarez impact hypothesis (for the K-T event in particular, but also as a candidate as a more general theory of extinction), and the modern biodiversity crisis, with its threat of contemporary mass extinction. Though there have always been some paleontologists in each generation studying extinction per se (Norman D. Newell of the American Museum of Natural History kept the issue alive virtually single-handedly through the mid-twentieth century), paleontologists have preferred to stress evolution over extinction—as if extinction is a mere setback to the real process of change in life: evolution. It is now clear that much of life's evolution would not have occurred *when* it occurred *in the manner* that it occurred had there not been a biotic turnover—a pulse of extinction—preceding the evolutionary activity.[2]

But my present concern is not evolution per se (even as an accompanying response to extinction), but rather a search for an understanding of the causes of mass extinction events. Except for the sporadic disappearance of individual species ("background extinction"—Jablonski, 1986), extinction is clearly a cross-genealogical, *ecological* phenomenon. Consequently, any serious candidate as a general theory of mass extinction must be ecological in nature. The Alvarez hypothesis is an excellent case in point.

2. *The Alvarez Impact Hypothesis of Mass Extinction*

Today, the most celebrated candidate for a general theory of mass extinction is the proposition that, periodically, Earth collides with an extraterrestrial object (or objects) of such large proportions that the direct consequences, primarily atmospheric ones, are so instantaneously profound as to disrupt the foundational food chains in both terrestrial and oceanic biomes. Extinction is seen as a rapid event unfolding on the scale of days, years, and decades—rather than the millennia and even millions of years more conventionally invoked by paleontologists seeking to explain large-scale extinction events.

The development of the impact hypothesis (Alvarez, Alvarez, Asaro, and Michel, 1980) provides an object lesson in the workings of contemporary science. The Alvarez team measured the amount of the rare-earth metallic element iridium in a thin, red clay layer that lies right at the Cretaceous-Tertiary boundary, in rocks exposed in the cliffs at Gubbio, Italy. They expected to find the customary minute amount of iridium, which would shed some light on the amount of time elapsed when the clay layer was being laid down as sediment. Instead they found iridium (and other rare-earth elements, such as platinum) in extraordinarily high concentrations—concentrations otherwise known only from certain meteorites (though volcanic activity has, in some instances, produced iridium in comparable concentrations).

The Alvarez team postulated that extraterrestrial impact was the direct, proximal cause of the great extinction event that separates the Mesozoic from the Cenozoic eras. Their scenario called for the ejection of particulate matter and vapors into the upper reaches of the atmosphere in such quantities that solar radiation would have been significantly blocked from reaching Earth's surface—drastically disrupting photosynthetic activity of both land plants and marine phytoplankton. Worldwide mass extinction ensued, as ecosystems literally collapsed from the base of the food chain on up through the trophic levels of each and every local ecosystem the world over.

Since its initial formulation, much data have accumulated in support of the Alvarez team's initial analysis. The iridium layer has been identified widely in many other geological sections, in both marine and nonmarine settings. Corroborative evidence, such as the presence of shocked quartz (formed under intense pressures and associated with other, well-documented impact events), traces of wild-fires at the boundary, identification of the likely crater caused by the impact, and the careful documentation of many species disappearing right at the very iridium layer at the Cretaceous-Tertiary boundary, make a very strong case for the extraterrestrial impact hypothesis—though some geologists continue to maintain that the iridium layer and the entire causal chain of the extinction event are more likely the product of widespread volcanism as the Cretaceous came to a close.

The suggestion by paleontologists David M. Raup and J. J. Sepkoski, Jr. (1984) that mass extinctions demonstrate a regular periodicity of some 26 million years fits well with the Alvarez hypothesis and helped transform the notion from a causal explanation of an isolated event into a plausible general theory of extinction. For if mass extinctions come with clocklike regularity, by far the most likely source of regularity would be astronomical, where orbits of celestial bodies display such regularities at such orders of temporal magnitude. Attention shifted to the Oort Cloud, a hypothetical cluster of comets at the outer reaches of the solar system. One (or more) comets quickly replaced a meteor as the likely culprit, and the search was on for the cause of the periodic perturbation of the Oort Cloud that would send a volley of comets toward the sun in sufficient density that a few would almost without doubt intersect Earth's orbit, colliding and causing ecological pandemonium every 26 million years. The existence of an unknown tenth planet (Planet X!) was postulated, as was Nemesis (the Death Star), a putative sister star to the Sun, with which it forms a binary double. The putative existence of Planet X and Nemesis hinges on hypothetical perturbations of the Oort Cloud, itself a hypothetical construct. That there have indeed been such perturbations itself hinges on the claim of rhythmic periodicity of mass extinctions caused by extraterrestrial impact. All told, quite an elaborate scheme!

To date, there is no corroborative empirical evidence to support the existence of either the Oort Cloud, or Nemesis, or Planet X. Even the periodicity of extinction is under intense scrutiny, with mathematicians and paleontologists arguing that the data do not support the claim of periodicity. Though by no means decisively falsified, periodicity remains undemonstrated to the clear satisfaction of many, perhaps even most, geologists and paleontologists.

If the Alvarez model is to serve as a general model (whether through notions of periodicity or not), there are obvious predictions to be tested right here on Earth. Despite initial claims, for example, no one has yet found iridium layers at any of the other major extinction boundaries. Moreover, paleontologists generally see patterns of ecosystem decline beginning long before the actual events defining geological boundaries take place—even in the celebrated case of the Cretaceous-Tertiary extinctions, which saw many species drop out just at the iridium layer. Diversity within higher taxa as a rule declines for some period of time prior to the actual boundary. "Gradual" and "abrupt" are loaded terms, even in geological circles: ecosystem collapse taking place on the order of tens and hundreds of thousands of years is still relatively "abrupt" compared with the far longer periods of ecosystem stability preceding the extinction event. But it is "gradual" when compared with the Alvarez scenario of days, years and decades. The Cretaceous-Tertiary extinction patterns actually display elements of both long-term ecosystem decline and abrupt loss of species right at the immediate boundary.[3] The other mass extinctions represent loss of species over longer periods of time.

Finally, global extinction events are not wholly anomalous; rather, they occupy the extreme position on a spectrum going from more frequent local ecosystem collapse to massive cross-genealogical extinction events. The relatively finer, more regionalized subdivisions of geological time also represent turnovers in the fauna—much of it marked by significant extinction. No one supposes that such lesser episodes of extinction are caused by more local and less devastating extraterrestrial impact events. The conclusion seems unavoidable that, whatever the role played by extraterrestrial factors on some isolated extinction events, the search for a general ecological theory of extinction must refocus its gaze on earthbound causes.

3. *Earthbound Causes of Mass Extinction*

For over a century preceding the Alvarez proposal, geologists argued over two basic alternatives as root causes for mass extinction: global cooling versus massive habitat destruction. Much of the debate centered around events at the greatest of all mass extinctions: that occurring at the Permo-Triassic boundary some 245 million years ago.

Most of the fossil record consists of the remains of marine invertebrates that lived in shallow water seas that have covered much of the continental surfaces for most of the past 600 million years.[4] The very end of the Permian was marked by widespread disappearance of normal-salinity marine conditions in many areas of the world, and paleontologists have long attributed the Permo-Triassic crisis in marine life simply to widespread habitat loss (Newell, 1963). Schopf (1974) reanalyzed the Permian extinctions in light of the theory of island biogeography, which implies certain reductions in diversity related to reduced sizes of regional habitat areas. The high-salinity evaporite deposits, coupled with the apparent retreat of widespread carboniferous glaciation, seemed to imply warm, arid conditions in the Upper Permian—a further argument that habitat loss, not global climate change (especially global temperature drop) was the root cause of the Permo-Triassic extinctions.

Yet geologists have long linked extinction with global cooling—with the relatively recent Ice Age (Pleistocene) events uppermost in mind, but with other clear signals from the geological past as well. The global cooling hypothesis received a strong boost when it was discovered that extensive southern hemisphere glaciation had occurred in the uppermost Ordovician Period, coinciding with a mass extinction then perhaps second in magnitude only to the Permo-Triassic event. Global climatic cooling seems associated with nearly all mass extinctions for which there is any definitive evidence at all (Donovan, 1989). Even the end of the Permian (undoubtedly arid, but not thereby necessarily warm) has recently yielded some signs of global cooling, as has the end of the Cretaceous (expected also under the Alvarez model).

I have recently argued that the species-area (habitat destruction/reduction) and global climate cooling hypotheses are actually complementary (1991) and, indeed, are merely two different aspects of the same basic, although complex, phenomenon. Most documented habitat alterations, such as for the Permo-Triassic and Cretaceous-Tertiary boundaries, involve the reduction in area of shallow-water epicontinental sea-ways. There are only two generally recognized mechanisms for signifi-cantly altering global sea level change: growth of polar ice caps and relative movement of crustal or tectonic plates, which may displace water onto, or drain water from, continents.

Thus the outline of a general theory of mass extinction is emerging: global climate change (often, though not necessarily, involving global temperature drop) is the usual underlying culprit. Climate change causes mass extinction in two basic ways. Stanley (1988) has convincingly shown that mass extinctions invariably involve greater devastation in the tropics than in the higher latitudes. Of course, species diversity is higher in the tropics. Higher latitude species are accustomed to wide fluctuations in temperature and tend to occupy relatively broad geographic areas. Tropi-cal species, in contrast, are far more narrowly focused on the relatively more minute environmental nuances they generally encounter, and trop-ical species accordingly tend to be far less widely dispersed than higher latitude species. Thus tropical species are especially vulnerable to global temperature drop. They are not accustomed to seasonal temperature change; there will be no warmer climes to which they can retreat (higher latitude species typically retreat toward the tropics in times of global tem-perature drop); and, with their smaller geographic areas (hence total pop-ulation sizes) tropical species are more vulnerable to extinction than higher latitude species. Colder times present a direct threat to tropical ecosystems.

The second, and more fundamental, way in which global climate change (cooling or warming) causes mass extinction is through the dis-ruption or outright destruction (as in the marine examples cited previ-ously) of accustomed habitat. Species survive as long as their component organisms continue to find suitable, "recognizable" habitat. Geologists have long known that environmental change is inevitable. What is *not* in-evitable is the long-held Darwinian supposition that evolutionary change will always follow environmental change. Species track environmental change: as long as habitats remain recognizable and *reachable* (given the differential abilities of organisms to relocate), species tend to persist unchanged. Failure to keep up, or simply to locate, such recognizable, ac-customed habitat results in extinction, not evolution, as the overwhelm-ing rule.[5]

Habitat disruption, rearrangement, and destruction occurs in the natural world chiefly through global climate change—whether this is a matter of direct loss of seaway, the dramatic migration southward of vegetation zones in front of advancing continental glaciers, or the more subtle (and perhaps less globally devastating) changes in terrestrial biomes, or major ecological regions, occasioned by changes in temperature, rainfall, and other environmental factors. Put the other way, global climate change—reflecting an interplay between amount of solar radiation reaching Earth and the topographic configuration of major features of Earth's surface—is the root cause of extinction, causing as it does massive, widespread habitat destruction and rearrangement. Global climate change is the ultimate, and habitat disruption the more proximal, cause of mass extinctions of the geological past.

We are now in a position to examine the changing position of hominid species in nature, the effects of climate change on patterns of human evolution, and the extent to which the actions of our own species, *Homo sapiens*, mimic the climate-induced loss of habitat that underlay mass extinctions of the ancient geological past.

4. *Hominid Species in Nature*

Species of sexually reproducing organisms typically develop discontinuous ranges. Local populations of any given species are integrated with local populations of many other species into local ecosystems. Such local populations, rather than the entire species as such, have niches and play their roles in ecosystems.[6] Early species of hominids,[7] like the living great ape species, share with most other known species this integration of local populations into local ecosystems. Indeed, a few local peoples within our own species, such as the Mbuti of Zaire, as described by Turnbull (1985), retain this same basic relation to nature—a mode of life that, in all likelihood, will unfortunately not be possible for them to pursue much longer.

That our early ancestral species occupied a structural place, and played a functional role, in nature like other species is strikingly borne home by the patterns of species origination and extinction among them. Paleontologist E. S. Vrba (1988) has documented a particularly crucial event affecting the regional ecology of eastern Africa 2.5 million years ago. Though the Pleistocene ("Ice Age") began 1.65 million years ago with the first of the four well-documented pulses of continental glaciation, a major cooling event occurred nearly a million years earlier.

A wide assortment of mammal species disappeared just below an abrupt break in the fauna, to be replaced later by a different group of species. The cooling pulse caused a radical reorganization, with forests and woodlands suddenly giving way to open savannahs with scattered clusters of trees. So far as the mammals were concerned, the habitats were altered radically and apparently rather suddenly. Species disappeared either through tracking familiar forest habitats elsewhere, or through actual extinction. The species that succeeded them migrated in from elsewhere, or were newly evolved. A "turnover pulse hypothesis" recognizes the several possible ways in which both disappearance of older, and appearance of newer, species may take place, suddenly and in concert, as a response to such environmental changes (Vrba 1985). The particularly compelling aspect of this account is that the factors underlying species extinction—namely, habitat disruption, fragmentation, and loss—are the very same as those conventionally cited as causes of speciation. Thus the causes of extinction may also serve as the very wellspring of the evolution of new species.

The mammal species at that time included antelopes, pigs, rhinos, and many other mammalian higher taxa—most certainly including hominids. The gracile australopithecines (specifically *Australopithecus africanus*) appear to have become extinct about 2.5 million years ago. Shortly thereafter, the earliest known species of each of two critically distinct hominid sublineages appeared: the first of the robust australopithecines (obligate vegetarians, with three or four related species spanning a million years) and a species, still poorly known, that belonged to our own sublineage, *Homo habilis*, "handy man," the earliest hominid species definitely to be associated with stone tools. That our early ancestral species showed patterns of both evolution and extinction very much in concert with many other contemporaneous mammal species eloquently testifies to the similar nature of the integration of all these species into their local ecosystems.

Though nothing much is known of the history of *Homo habilis*, its successor, *Homo erectus*, occupies a pivotal position in human evolutionary history. An enormously successful species, *H. erectus* appeared about 1.5 million years ago in East Africa—right after the onset of the first of the four major glaciation pulses of the Pleistocene epoch.[8] Thus the pattern holds: *Homo erectus* appears to have evolved in conjunction with a major event of climate change.

But, from then on, the picture changes in subtle but highly significant ways. Another great glacial pulse occurred about .9 million years ago, and once again we see a correlated response in *Homo erectus*. But this time the response is neither evolution nor extinction, but is, rather, "ecogeographic": *Homo erectus* shows up in Eurasia for the first time .9 million years ago, migrating out of Africa *northward* into the very teeth of the northern hemisphere glacial pulse. Now we begin to detect something novel in Earth's history, a species that can occupy many different environments, because it can remake its environments, owing to the advent of culture.

Advances in culture undoubtedly underlay the increasing ecological flexibility of *Homo erectus* over its forebears. The earliest traces of the use of fire date back to 1.5 million years at *Homo erectus* sites in east Africa. The species spread far and wide over the old world, leaving its bones— and, more commonly, its stone tools, including the famous stone axes—at archeological sites from Cape Town north to France and east to Beijing and Java. Prior to that, *Homo erectus* had already survived for 1.3 million years, according to many paleoanthropologists (Rightmire, 1990), without significant evolutionary, or even cultural, change. In so doing, the species literally weathered many minor, and several major, climatic oscillations: surviving, yet not evolving to any significant degree.

But now there is something new. If evolutionary stasis is the normal outcome of continued habitat recognition, now, with this remarkable expansion, we can only conclude that *H. erectus* significantly broadened the hominid concept of "suitable habitat." It did so through the development of an advanced culture, a unique evolutionary specialization that, ironically if not paradoxically, conferred greatly expanded ecological flexibility and generality on this species and its descendants—most certainly including ourselves (Eldredge and Tattersall, 1982).

It is difficult to specify the cause of the demise of *H. erectus* after a successful ecological "run" of nearly a million and a half years. Certainly no event of climatic change is implicated. Hominid fossils of 300,000-year-old vintage are not common. They tend to occur as isolated skulls, and some paleoanthropologists (Tattersall, 1986) suggest that they may represent several distinct species. If so, what caused so much evolutionary activity (speciation), and how the appearance of the new species affected the ancestral *Homo erectus*, remain obscure.

Our own species, *Homo sapiens*, is inferred to have arisen in Africa some 150,000 to 200,000 years ago on the basis of present-day diversity of mitochondrial DNA sequences (Cann, Stoneking, and Wilson, 1987). The molecular analysis agrees in general with the fossil record: the earliest "anatomically modern" humans are African (Stringer, 1988). They (or rather, "we") were present in the Mideast by 90,000 years ago. By 35,000 years ago, we were widespread in Europe. Therein lies the first real (albeit circumstantial) evidence of *Homo sapiens* implicated with an extinction event, perhaps for the first time as a causal agent instead of the previous historical role of victim.

Neanderthals (a distinct species of hominid according to an emerging majority of paleoanthropologists) had occupied Europe beginning roughly 100,000 years ago. After the arrival of *Homo sapiens* in Europe, there was a brief overlap period lasting several thousands of years. Thereafter, the neanderthals disappeared and we, of course, remained. It is difficult to avoid the conclusion that our arrival had something to do with the extinction of neanderthals, though whether the extinction occurred through direct negative interaction (meaning warfare) or the more subtle actions of competitive exclusion is impossible to say.[9]

Paleomammalogist Ross D. MacPhee and colleagues (Burney and MacPhee, 1988) have discussed the close correlation between the spread of *Homo sapiens* throughout the world and the extinction of many species, particularly of large mammal game animals. Such events occurred 30 to 40,000 years ago in Australia, 15 to 20,000 years ago in North America, and more recently as humans reached larger islands off continental coasts. Anthropologist Paul Martin (Martin, 1984) has devoted years to the development of his "Pleistocene Overkill" (or "Blitzkrieg") hypothesis, which is based on detailed studies of the arrival of humans in North America and the extinction of many Pleistocene species.

There are some cautions, however, to the full-scale acceptance of the notion that humans single-handedly have caused the extinction of many species, particularly on first arrival, by overhunting. *Homo sapiens* evolved in Africa, yet the Pleistocene large mammal fauna remain more intact there than anywhere else. And small mammals and other, nonmammal species that were presumably not high on the hunting list also became extinct in many instances at about the same time as the large mammals. The evidence in Madagascar requires particular attention. Tempting as it is to blame the disappearance of larger animals in the fauna purely on the arrival of humans in Madagascar (a scant 2,000 years ago), climatic events have apparently also been at work in comparatively recent times to modify the forest habitat (Burney and MacPhee, 1988). The issue is crucial, as human destruction of forest habitat in Madagascar directly threatens the continued existence of many additional species. But hunting as such, while undoubtedly causing the extinction of many species, seems to be only a partial explanation for the many extinctions of the past 20,000 years or so.

It was the invention of agriculture, I have concluded, that brought *Homo sapiens* solidly into the extinction business in a truly global and all-encompassing way. Though agriculture, of course, originated independently in a number of different areas, the oldest evidence for cultivation of plants appears to be within the Natufian culture of the Middle East of approximately 11,000 years ago. Agriculture transforms terrestrial habitats into monocultural expanses. Its invention both demanded and enabled a relatively settled existence which, coupled with a ready and dependable food supply, soon led to unprecedented human population expansion. Population growth fed back the need for additional food production, which (at least in pre-industrial days) usually resulted in expanding the extent of the areas under cultivation.[10]

Through agriculture (including tree harvesting as well as the destruction of forests for cultivating crops), humans have assumed the role formerly played exclusively by global climate change—or by the odd collision between Earth and extraterrestrial bodies. The tropics are especially vulnerable. Greater numbers of species, with, on average, much narrower geographic ranges exist in the tropics than in the higher latitudes. Transformation of forest ecosystems in the Third World has every bit the same relative impact on those systems as have events of global climatic change in the geological past.[11]

Agriculture has transformed the very stance that our species holds vis à vis nature. Neither agricultural communities nor the settlements they support bear the same relation to local ecosystems that had been traditionally held by local populations of our own, all other hominid, and, indeed, all other species whatever, since the dawn of complex life on Earth. We tend not to see ourselves in and of nature, but rather as conquerors of nature. We seek the Biblically enjoined "dominion" over nature and measure progress in large terms by the addition of still more layers of insulation between ourselves and the natural world. Nor is this surprising; it is consistent with the deepest roots of our history, going back at least 1.5 million years with the origin of *Homo erectus*. But the difference is not entirely one of degree: until fairly recently, both despite and because of our cultural innovations, we have continued to interact with local ecosystems at the (local) population level.

That such is no longer the case reflects the global economic interconnectedness of our species. The pattern developed as soon as it became more important to interact with other people living elsewhere than with other species occurring locally. Abetted by a cocoonlike sense of shielding from nature, we are the first species whose organisms really seem to act outside of nature. We are a global species, not one whose components are any longer integral parts of most of the local ecosystems of which we used to be a part.

But we have not really and truly escaped nature, as the headlines about the hole in the ozone layer affirm in the morning newspaper. Species, in my view, are historical entities, packages of genetic information with beginnings, histories, and ends. Species are not economic entities per se, interacting as parts of some form of biological economic (i.e., ecological) system, as has frequently been assumed by evolutionary biologists (Eldredge, 1985b, 1986, 1989). Yet our species, having abandoned the traditional ecological role of local populations that are integrated into local ecosystems, is still very much part of nature. We are a global species, interacting among ourselves globally, and interacting with the biosphere now on global terms. In this, we appear to violate the generalization that species per se cannot be economic interactors. But social systems, certainly including human social systems, are amalgams of economic and reproductive functions (Eldredge and Grene, 1992). Our global interactivity is essentially economic and specieswide with the result that we are still within, and interacting with, nature. But, unlike any other species, we are doing so on a global, rather than local, scale.

Thus it is a mistake to think that we are above and beyond nature. Though we still seem to be very much at odds (almost at war) with nature, the truth is that we are still very much dependent upon the global Gaian atmospheric, oceanic, and hydrological cycles—the very basic qualities of air and water that our technological effluvia are now beginning to attack so alarmingly. For this pragmatic reason, if no other, we must shed this illusion of separateness from nature—an illusion that springs from our shift away from being parts of local ecosystems and springs as well from the comforts afforded by a collectively very clever mass of humanity. We cannot try to recapture the past and once again rejoin local ecosystems (nor would I wish to do so myself even were it possible). But, because we depend on global biospheric health for our own continued existence, we must take the adage, "Think globally, act locally" seriously. Global biospheric health adds up from and depends on the health of local ecosystems. We must strive to put a halt to continued habitat destruction, meaning, as the bottom line, that we must curb global human population growth. Only if enough ecosystems are left in, or restored to, a sound functioning state can the biosphere persist.

Therein lies our own fate. It would be ironic, indeed, if the very strategy of divorce from local ecosystems that we have followed so closely for so long now turns out to be our ultimate undoing. Thus I return to human responsibility: responsibility, if to no others, at least to ourselves, our own species. As rational creatures, pondering the past, present, and future of life, we may suppose that the biosphere will persist and, indeed, flourish, were our species to become extinct. But as a living, functioning species of this planet, our destructive and ultimately suicidal behavior ought to be resisted just as strenuously as society morally opposes most forms of individual suicide. It is our job to keep going, indeed to thrive. But we can no longer do so at the expense of the rest of the biosphere. Curbing human population growth—assuming that it is not too late—will stave off massive biospheric destruction. Therein lies our own major hope for continued existence.

References

Alvarez, Luis W.; Walter Alvarez; Frank Asaro; and Helen V. Michel. 1980. "Extraterrestrial cause for the Cretaceous-Tertiary extinction." *Science* 208:1095–1108.

Burney, D. A., and R. D. E. MacPhee. 1988. "Mysterious island." *Natural History* 7/88:47–55.

Cann, R. L.; M. Stoneking; and A. C. Wilson. 1987. "Mitochondrial DNA and human evolution." *Nature* 325:31–36.

Caporael, L. R.; R. M. Dawes; J. M. Orbell; and A. J. C. van de Kragt. 1989. "Selfishness examined: Cooperation in the absence of egoistic incentives." *Behavioral and Brain Sciences* 12:683–739.

Cuvier, Georges. 1812/1831. *A Discourse on the Revolutions of the Surface of the Globe, and the Changes Thereby Produced in the Animal Kingdom*. Philadelphia: Carey and Lea (1831; translation of French edition of 1812).

Donovan, S. K., ed. 1989. *Mass Extinctions: Processes and Evidence*. New York: Columbia University Press.

Ehrlich, Paul R., and Edward O. Wilson. 1991. "Biodiversity studies: science and policy." *Science* 253:758–762.

Eldredge, Niles. 1985a. *Time Frames*. New York: Simon and Schuster. Reprint edition. 1989, Princeton: Princeton University Press.

Eldredge, Niles. 1985b. *Unfinished Synthesis: Biological Hierarchies and Modern Evolutionary Thought*. New York: Oxford University Press.

Eldredge, Niles. 1986. "Information, economics and evolution." *Annual Reviews of Ecology and Systematics* 17:351–369.

Eldredge, Niles. 1989. *Macroevolutionary Dynamics: Species, Niches and Adaptive Peaks*. New York: McGraw-Hill.

Eldredge, Niles. 1991. *The Miner's Canary. Unraveling the Mysteries of Extinction*. New York: Prentice Hall (Simon and Schuster).

Eldredge, Niles, and Marjorie Grene. 1992. *Interactions: The Biological Context of Social Systems.* New York: Columbia University Press.

Eldredge, Niles, and Ian Tattersall. 1982. *The Myths of Human Evolution.* New York: Columbia University Press.

Halstead, L. B. 1990. "Cretaceous-Tertiary (Terrestrial)." In D. E. G. Briggs and P. R. Crowther, eds., *Palaeobiology: A Synthesis* (pp. 203–207). Oxford: Blackwell Scientific Publications.

Jablonski, David. 1986. "Background and mass extinctions: The alteration of macroevolutionary regimes." *Science* 231:129–133.

Lovelock, James E. 1987. *Gaia: A New Look at Life on Earth.* Oxford: Oxford University Press.

Mann, C. C. 1991. "Extinction: Are ecologists crying wolf?" *Science* 253:736–738.

Martin, P. S. 1984. "Catastrophic extinctions and Late Pleistocene blitzkrieg: Two radiocarbon tests." In M. H. Nitecki, ed., *Extinctions* (pp. 153–189). Chicago: University of Chicago Press.

Newell, Norman D. 1963. "Crises in the history of life." *Scientific American* 208 (No. 2, February):76–92.

Raup, David M. 1979. "Size of the Permo-Triassic bottleneck and its evolutionary implications." *Science* 206:217–218.

Raup, David M., and John J. Sepkoski, Jr. 1984. "Periodicity of extinctions in the geologic past." *Proceedings of the National Academy of Sciences, U. S. A.* 81:801–805.

Rightmire, G. P. 1990. *The Evolution of Homo Erectus: Comparative Anatomical Studies of an Extinct Human Species.* Cambridge: Cambridge University Press.

Rolston, Holmes III. 1988. *Environmental Ethics: Duties to and Values in the Natural World.* Philadelphia: Temple University Press.

Schopf, T. J. M. 1974. "Permo-Triassic extinction: Relation to sea-floor spreading." *Journal of Geology* 82:129–43.

Sheehan, P. M.; D. E. Fastovsky; R. G. Hoffmann; C. B. Berghaus; and D. L. Gabriel. 1991. "Sudden extinction of the dinosaurs: Latest Cretaceous, upper great plains, U. S. A." *Science* 254:835–839.

Soulé, Michael E., ed. 1986. *Conservation Biology: The Science of Scarcity and Diversity.* Sunderland, MA: Sinauer Associates.

Stanley, Steven, M. 1988. "Paleozoic mass extinctions: Shared patterns suggest global cooling as a common cause." *American Journal of Science* 288:334–352.

Stringer, C. B. 1988. "*Homo sapiens.*" In Ian Tattersall, E. Delson, and J. van Couvering, eds., *Encyclopedia of Human Evolution* (pp. 267–274). New York: Garland Publishing.

Tattersall, I. 1986. "Species recognition in human paleontology." *Journal of Human Evolution* 15:165–176.

Turnbull, C. M. 1985. "Cultural loss can foreshadow human extinctions: The influence of modern civilization." Pages 175–92 in Hoage, R. J., ed., *Animal Extinctions: What Everyone Should Know*. Washington, D.C.: Smithsonian Institution Press.

Vrba, E. S. 1985. "Environment and evolution: Alternative causes of the temporal distribution of evolutionary events." *South African Journal of Science* 81:229–236.

Vrba, E. S. 1988. "Late Pliocene climatic events and hominid evolution." In F. E. Grine, ed., *The Evolutionary History of the Robust Australopithecines*. New York: Aldine de Gruyter.

Western, David, and Mary Pearl, eds. 1989. *Conservation for the Twenty-first Century*. New York: Oxford University Press.

Wilson, Edward O., ed. 1988. *Biodiversity*. Washington, D.C.: National Academy Press.

Endnotes

1. Most early geologists who developed the geological time scale were (to the extent that their religious views are known) "creationists," in the sense that they saw no essential conflict between their Judeo-Christian religious traditions and their geological conclusions. Recent claims by creationists that the geological time scale was fabricated by evolution-minded geologists anxious to demonstrate continuity and progress in the history of life as a validation of the very notion of evolution are utterly false. The basic outline of the geological time scale was the work of creationists!

2. Some major groups, such as the teleost fishes, a group of bony fishes including most living species, and the angiosperm or flowering plants arose and diversified without apparent prior extinction of ecologically similar groups. The earliest land plants and animals evolved through invasion of habitat not previously occupied by ecological forerunners. But much of large-scale macroevolution follows the disappearance of taxa, typically after long histories, with their replacement by groups that radiate only after their disappearance. Mammals and dinosaurs arose at about the same time (Upper Triassic), but only the dinosaurs diversified significantly. Only after the final demise of the dinosaurs (and allied aquatic and flying reptilian taxa) did mammals then radiate and come—very quickly—to be the dominant vertebrate components in most of the world's terrestrial ecosystems.

3. Paleontological discussions of abrupt versus gradual patterns of extinction are often at cross purposes. For example, Halstead (1990) has discussed the gradual decline of diversity of several groups, including dinosaurs, over the course of several millions of years preceding the Cretaecous-Tertiary boundary. Sheehan et al. (1991), examining a shorter temporal interval than that discussed by Halstead, dispute gradual loss of dinosaur species as the Creta-

ceous grew to a close. Yet the two patterns are not mutually exclusive: the data indeed suggest that (a) there was an abrupt loss of Cretaceous species (often in several distinct waves) coinciding with (or nearly so) the iridium boundary layer, but that (b) many taxa in fact do display a decline in species-level diversity for several millions of years prior to the abrupt event.

4. The polar ice caps currently tie up so much water that the continents today stand anomalously high and dry.

5. I discuss these generalizations in greater detail, with examples and documentation, elsewhere, especially Eldredge, 1985a, 1989, and 1991.

6. I have discussed ecological and genealogical systems generally, and the ontological nature of species in particular, in Eldredge. 1985b, 1986, and 1989.

7. Systematists include both the great apes and our own species, *Homo sapiens*, in the single Family Hominidae. However, my informal usage of "hominid" refers strictly to our human species and those fossil species of our direct lineage.

8. We may eventually learn that *Homo erectus* evolved earlier than present evidence indicates. Nonetheless, given the intensity of the search for hominid fossils, especially over the past thirty years, the human fossil record has actually grown to a respectable density, and the emerging patterns are less likely to be changed radically in the future than in the earlier days of human paleontology. Famous last words!

9. That the neanderthals became extinct because they hybridized with newly arrived populations of anatomically modern *Homo sapiens* is highly unlikely. Hybridizing leading to extinction of well-differentiated taxa occupying such large territories is not commonly even hypothesized, let alone documented, in the biological literature. Neanderthals and modern humans were quite distinct biological species, according to prevailing opinion—meaning that interbreeding between the two was in all likelihood impossible.

10. A recent report (*The New York Times*, November 3, 1991, Sec. 4, p. 2) indicates that, at least in the U.S., less land is under cultivation to support more people than was the case only 60 years ago—the result of greater agricultural efficiency and the invention of the internal combustion engine, which replaced grain-consuming draught animals for transportation.

11. The disparity between the "haves" of the industrialized, predominantly higher latitude nations and the "have nots" of the tropical Third World raises important ethical and pragmatic issues that must be addressed, lest the charge of hypocrisy against conservationists from the developed nations be made to stick and efforts to save tropical ecosystems thwarted.

Evolutionary Ethics
A Defense

MICHAEL RUSE

▪ ▪ ▪ *Editor's Introduction*

Michael Ruse is the most celebrated philosopher in the world for his untiring effort to join biology and ethics. "The question is not whether biology—specifically, our evolution—is connected with ethics, but how" (p. 93). "Our moral sense is an adaptation helping us in the struggle for existence and reproduction, no less than hands and eyes, teeth and feet. It is a cost-effective way of getting us to cooperate" (p. 97). Literal, moral altruism is produced in the service of a biological, technical "altruism" that selects for cooperation out of enlightened genetic self-interest.

Ruse knows that the philosophical tradition has a bad taste for deriving biology from ethics. Justifiably so, Ruse concedes, for he too disapproves of past efforts. Misunderstanding the implications of Darwinism for ethics, earlier evolutionary ethics extrapolated the struggle in nature into culture and was too combative. Rather, evolution inclines humans to cooperation. "In order to achieve 'altruism,' we are altruistic! To make us cooperate for our biological ends, evolution has filled us full of thoughts about right and wrong, about the need to help our fellows, and so forth. . . . It is in our biological interests to cooperate. Thus we have evolved innate mental dispositions . . . inclining us to cooperate, in the name of this thing which we call 'morality' " (pp. 96–97).

Evolution has also filled us with thoughts about truth—a problematic complication, however, because this "truth" is illusion. Our self-interests are powerful drives, deeply ingrained from our animal heritage; and they must now be offset, moderated, if we are to succeed in a cooperative culture. The selfish genes dispose us to form a belief that we ought to act altruistically because this is some kind of objective truth, transcending human affairs. We come to believe falsely but necessarily that moral truth has what Ruse calls "an objective reference" (p. 101). Metaphysics and theology authorize morality. In fact, both metaphysics and theology are illusions—like ouija boards. In fact, we are acting in our self-interests, deceived into thinking we are and ought to be genuinely altruistic. Literal moral altruism is illusory altruism, serving biological self-interests while pretending it does not.

One might think that, once the secret is out, the whole game of pretense will unravel. Ethics will collapse. Not so. Ruse insists that the pretense can and ought to continue, for it is not really pretense in a less pretentious sense. Rather, it is quite true so far as it claims that a cooperative morality is the best thing for humans who try to live together in culture. However, it is probably better to think of such a claim as *reasonable* rather than as *true* fact. For Ruse does wish to continue the cooperative ethic written into our genes, manifest in the "epigenetic rules."

Such an ethic is remarkably similar to what moral philosophers themselves recommend, notably John Rawls, whose *A Theory of Justice* is probably the most influential work in moral philosophy in recent decades. However, in Rawls, there is no illusion; the choices he urges us to make are free and rational; we are one among thoughtful, self-interested equals. In Ruse, there is illusion. Morality "is just an illusion, fobbed off on us to promote 'altruism' " (p. 100).

But if there seems too much illusion, we can put it another way: "What is really important to the evolutionist's case is the claim that ethics is illusory inasmuch as it persuades us that it has an objective reference" (p. 101). Ethics is not illusory in the sense that it doesn't work, nor that it is not really the best thing for humans, nor in the sense that people who act ethically are out of touch with the world they need to live in. Ethics does work; people who act ethically are in excellent touch with the world they need to live in. The illusion is that there is any "objective reference" to ethics. The fact is that ethics is evolution's way of making a more humane form of life.

Nevertheless, curiously, ethics will not work, certainly not very well, without such an illusion. "There are good biological reasons why it is part of our nature to objectify morality. If we did not regard it as binding, we would ignore it. It is precisely because we think that morality is more than mere subjective desires, that we are led to obey it." "We have to believe in morality, otherwise it will not work" (p. 102). Later, Langdon Gilkey is going to press other sociobiologists, as well as Ruse, with *ad hominem* questions. If ethics is an illusion, how are sociobiologists going to find the will to continue in the illusion? Or to break out of it to something higher?

Is all seemingly altruistic conduct really enlightened self-interest? Must we continue an illusion otherwise for such altruism to be effective? Is such altruism always a means to producing more offspring in the next generation? What, then, is the role of reason in determining what we ought to do? Will free and rational persons choose the kind of morality that, according to Ruse, we discover ourselves to have biologically? What does it mean for morality to have "objective reference"? Is morality natural? Does morality follow from human nature? Are the rules different in culture and in nature, so that, for instance, natural selection might be relaxed or transcended in ethics? Does action that is just and fair coincide, always or usually, with what is in our enlightened self-interest? Ruse not only raises the central issues; he sets the stage for the debates to follow about biology and ethics.

Michael Ruse is professor of philosophy at the University of Guelph and the author of fourteen books on the philosophy of biology. Among them are: *The Philosophy of Biology* (1973), *Sociobiology: Sense or Nonsense?* (1979), *The Darwinian Revolution: Science Red in Tooth and Claw* (1979), *Darwinism Defended: A Guide to the Evolution Controversies* (1982), *Taking Darwin Seriously: A Naturalistic Approach to Philosophy* (1986), *Philosophy of Biology Today* (1988), and *The Darwinian Paradigm* (1989). His books have been translated into Russian, Italian, Spanish, Portuguese, and Polish. He has also written over four dozen articles in academic journals, four dozen chapters in anthologies and books, and is the founding editor of the journal *Biology and Philosophy*. He is a Fellow of the Royal Society of Canada and Fellow of the American Association for the Advancement of Science.

The time has come to take seriously the fact that we humans are modified monkeys, not the favored Creation of a Benevolent God on the Sixth Day. In particular, we must recognize our biological past in trying to understand our interactions with others. We must think again especially about our so-called "ethical principles." The question is not whether biology—specifically, our evolution—is connected with ethics, but how. Thanks to recent developments in biological science, we can now throw considerable light on this problem.

I begin with a brief historical introduction. Then, I move to the core of my scientific and philosophical case. I conclude by taking up some central objections.

1. *Traditional Evolutionary Ethics*

Charles Darwin argued that all organisms (including ourselves) originated through a slow, natural process of evolution. Also, Darwin suggested a mechanism. More organisms are born than can survive and reproduce. This leads to competition. The winners are thus "naturally selected," and hence change ensues in the direction of increased "adaptiveness." It is hardly true that Darwin, or even science generally, brought about the death of Christianity; but, after the *Origin* increasing numbers of people began to turn from the Bible and to look toward evolution, in some form, for moral insight and guidance (Ruse, 1979a; Russett, 1976). The resulting evolutionary ethics was generally called "social Darwinism," although, as many have noted, it owed its genesis more to that general man of Victorian science, Herbert Spencer, than to Charles Darwin himself.

A full moral system needs two parts. The "substantive" or "normative" ethical component offers actual guidance: "Thou shalt not kill." The "metaethical" dimension offers foundations or justification, as in: "That which you should do is that which God wills." Without these two parts, any system is incomplete (Taylor, 1978).

To the social Darwinians, the metaethical foundations they sought lay readily at hand. They exist in the perceived nature of the evolutionary process. Supposedly, we have a progression, from simple to complex, from amoeba to man, from (as Spencer happily pointed out) savage to Englishman (Spencer, 1852, 1857). This progress is a good thing and conveys immediate worth. We need no further justification of what ought to be. And now, at once, we have the substantive directives of our system. Morally, we should aid and promote—and not hinder—the evolutionary process. Further, if, as was supposedly claimed by Darwin and certainly echoed by Spencer, the evolutionary process begins with a bloody struggle for existence and concludes with the triumph of the fittest, then so be it. Our obligation is to prize the strong and successful and let the weakest go to the wall (Ruse, 1985).

Of course, as many pointed out—most splendidly, Darwin's great supporter and ardent co-evolutionist, Thomas Henry Huxley (1901)—none of this will do. Evolution simply is not progressive (Williams, 1966). Evolutionary lines branch all over the place, making it quite impossible to offer true assessments of top and bottom, higher and lower, better and worse. Among today's organisms, venereal diseases thrive, whereas the great apes stand near extinction. Is gonorrhea really superior to the chimpanzee? And, following up the metaethical inadequacies, at the fundamental level, when it comes to the substance of the matter, if anything is false, social Darwinism is false. Morality does not consist in walking over the weak and the sick, the very young and the very old. Someone who tells you otherwise is an ethical cretin.

Social Darwinism (and, as many concluded, any kind of evolutionary ethics) is wrong—not just mistaken, but fundamentally misguided. Why? The answer was pinpointed by such philosophers as David Hume (in the eighteenth century) and G. E. Moore (in the twentieth century). Hume noted that you simply cannot go straight from talk of facts (like evolution) to talk of morals and obligations, from "is" language to "ought" language.

> In every system of morality, which I have hitherto met with, I have always remark'd, that the author proceeds for some time in the ordinary way of reasoning, and establishes the being of a God, or makes observations concerning human affairs; when of a sudden I am surpriz'd to find, that instead of the usual copulations of propositions, is, and is not, I meet with no proposition that is not connected with an ought, or and ought not. This change is imperceptible; but is, however, of the last consequence. For as this ought, or ought not, expresses some new relation or affirmation, 'tis the same time that a reason should be given, for what seems altogether inconceivable, how this new relation can be a deduction from others, which are entirely different from it. (Hume, [1739]1978, p.469)

In 1903, Moore backed up this point in his *Principia Ethica*, arguing that all who derive morality from the physical world stand convicted of the "naturalistic fallacy." Moore noted explicitly that evolutionary ethicists are major offenders, going from talk of the facts and process of evolution to tell of what one ought (or ought not) do.

So traditional evolutionary ethics ground to a complete stop. It promoted a grotesque distortion of true morality and could do so only because its foundations were rotten (Flew, 1967). And so matters have rested for three-quarters of a century. Now, however, the time has come for the case to be reopened. Let us see why.

2. *Morality as a Product of Evolution*

We must begin with biological science as this describes the evolution of the human moral capacity. In fact, as Charles Darwin pointed out, contrary to the Spencerian interpretations of the evolutionary process, although the process may start with competition for limited resources, and although there is a struggle for existence (more strictly, a struggle for reproduction), this certainly does not imply that there will always be fierce and ongoing hand-to-hand combat. Especially between members of the same species, much more personal benefit can frequently be achieved through cooperation—a kind of enlightened self-interest (Darwin, 1859, 1871). Thus, if I battle with another of my species until one is totally vanquished, no one really gains, for even the winner, beaten and exhausted, is left in poor shape for future tasks. By contrast, if we cooperate, although we must share the contested resource, there will be no big losers and both will benefit (Trivers, 1971; Wilson, 1975; Dawkins, 1976; Ruse, 1979b).

All such cooperation for personal evolutionary gain is known in a technical sense as *altruism*. I emphasize that this term is rooted in metaphor, even though now it also has the just-stipulated formal biological meaning. There is no implication that evolutionary "altruism" (working together for biological payoff) is inevitably associated with moral altruism (*altruism* in the original literal sense, implying a conscious being helping others because it is right and proper to do so). There is no more connection than between the technical definition of *work* in physics and what you and I do in the garden on Saturday afternoons, when we mow the grass.

However, just as mowing the lawn does involve "work" in the physicist's sense, those who today study the evolution of social behavior ("sociobiologists") argue that (literal, moral) altruism might be one way in which (biological, metaphorical) "altruism" could be achieved (Wilson, 1978; Ruse and Wilson, 1986). Further, they argue that in humans, and perhaps also in the great apes, such a possibility is a reality. Literal, moral altruism is a major way in which advantageous biological cooperation is achieved. Humans are the kinds of animals that benefit biologically from cooperation within their groups (biological altruism), and (literal, moral) altruism is the way in which we achieve that end (Lovejoy, 1981).

It was not inevitable that altruistic inclinations would develop as one of the human adaptations. Judging from what we know of ourselves and other animals, there were a number of other ways in which biological "altruism" might have been affected (Lumsden and Wilson, 1983). Humans could have gone the route of the ants. They are highly social, having taken biological "altruism" to its highest pitch, through what one might call "genetic hardwiring." Ants are machinelike, working in their nests according to innate dispositions, triggered by chemicals that produce behaviors mechanically ("pheromones") and the like (Wilson, 1971).

There are great biological advantages to this kind of functioning. It eliminates the need for learning, it cuts down on the mistakes, and much more. Unfortunately, however, this is all bought at the expense of any kind of flexibility. If circumstances change, individual ants cannot respond. This does not matter so much in the case of ants, since (biologically speaking) they are cheap to produce. Unfortunately, humans require significant biological investment, and so, apparently, the production of "altruism" through innate, unalterable forces, poses too much of a risk.

Since the ant-option is closed, we humans might theoretically have achieved "altruism" by going to the other extreme. We might have evolved super-brains, rationally calculating at each point a detailed cost-benefit analysis of whether a certain course of action is in our best interests. "Should I help you prepare for a difficult test? What's in it for me? Will you pay me? Do I need help in return? Or what?" Here, there is nothing more than a disinterested, though perhaps complex, calculation of personal benefits. However, we have clearly not evolved this way. Such a super-brain would have high biological costs and might not be that efficient. By the time I have decided whether or not to save the child from the speeding bus, the dreadful event has occurred (Lumsden and Wilson, 1981; Ruse and Wilson, 1986).

Human evolution seems rather to have been driven toward a middle-of-the-road position. In order to achieve "altruism," we are altruistic! To make us cooperate for our biological ends, evolution has filled us full of thoughts about right and wrong, about the need to help our fellows, and so forth. We are obviously not totally selfless. Of course, our normal disposition is to look after ourselves; we have to do that if we are to reproduce. However, it is in our biological interests to cooperate. Thus we have evolved innate mental dispositions (what the sociobiologists Charles Lumsden and Edward O. Wilson call "epigenetic rules") inclining us to cooperate, in the name of this thing we call "morality" (Lumsden and Wilson, 1981). We have no choice about this morality of which we are aware. We do not choose whether to have a conscience. But, unlike the ants, we can certainly choose whether or not to obey the dictates of our conscience. We are not blindly locked into our courses of action like robots. We are inclined to behave morally, but not predestined either to a particular moral policy or to full obedience to it.

This, then, is the Darwinian biologist's case for the evolution of morality. Our moral sense is an adaptation helping us in the struggle for existence and reproduction, no less than hands and eyes, teeth and feet. It is a cost-effective way of getting us to cooperate, which avoids both the pitfalls of blind action and the expense of a super-brain of pure rationality.

3. *Substantive Ethics*

But what has any of this to do with the questions that philosophers find pressing and interesting? Let us grant the scientific case, just sketched. What now of substantive ethics? If we think that what has just been said has any relevance to foundations, then surely we violate Hume's law and smash into the naturalistic fallacy, no less than does the Spencerian?

Turning first to the moral norms endorsed by the modern evolutionist, there is little to haunt us from the past. As we have just seen, the whole point of today's approach is that we transcend a bloody struggle for existence, both in thought and deed. Of course, humans are selfish and violent at times. This has been admitted. But, no less than the moralist, the evolutionist denies that this darker side to human beings has anything to do with moral urges. What excites the evolutionist is the fact that we have feelings of moral obligation laid over our brute biological nature, inclining us to be decent for altruistic reasons.

What is the actual content, the substance, of a modern evolutionary ethic? At this point, we turn to philosophers for guidance! After all, these are the people who intend to uncover the basic rules that govern our ethical lives. The evolutionist may modify or even reject the philosophers' claims; but, given the evolutionist's central empirical and theoretical hypothesis that normal, regular morality is what our biology uses to promote "altruism," the evolutionist can start with a presumption that the findings of the philosophers will tell much about what sort of morality has been biologically selected for.

There is no need for philosophers to be apprehensive here. Or for the biologists to worry either. Claims of some of today's leading thinkers sound almost as if they were prepared expressly to fill the evolutionist's bill—a point that the philosophical thinkers themselves have acknowledged. Consider the ideas of John Rawls, whose *A Theory of Justice* deservedly holds its place as the major work in moral philosophy of the last decades. Rawls writes:

> The guiding idea is that the principles of justice for the basic structure of society are . . . the principles that free and rational persons concerned to further their own interests would accept in an initial position of equality as defining the fundamental terms of their association. These principles are to regulate all further agreements; they specify the kinds of social cooperation that can be entered into and the forms of government that can be established. This way of regarding the principles of justice I shall call justice as fairness. (Rawls, 1971, p. 11)

How exactly does one spell out these principles that would be adopted by "free and rational persons concerned to further their interests"? Here, Rawls invites us to put ourselves behind a "veil of ignorance." If we knew we were going to be born into a society, and that we would be healthy, handsome, wise, and rich, we would opt for a system that favors the fortunate. But, we might be sick, ugly, stupid, and poor. Thus, in our ignorance, we will opt for a just society, governed by rules that would best benefit us, no matter what state or post we might have in that society.

Under these conditions, a just society is seen to be one which, first, maximizes liberty and freedom, and second, distributes society's rewards so that everyone benefits as much as possible. Rawls is not arguing for some kind of communistic, totally equal distribution of goods. Rather, the distribution must help the unfortunate as well as the fortunate. If you could show that the only way to get statewide good-quality medical care is by paying doctors twice as much as anyone else, then so be it.

I need hardly say how readily all of this meshes with the evolutionary approach. For both the biologist and the Rawlsian, the question is how one might obtain right action from groups of people whose natural inclination is (or rather, of whom one would expect the natural inclination to be) that of looking after themselves. And, in both cases, the answer is found in a form of enlightened self-interest. We behave morally because, ultimately, there is more in it for us than if we do not. Where the evolutionist goes beyond the Rawlsian is in linking the principles of justice to our biological past, via the epigenetic rules. This is a great bonus, for Rawls himself admits that his own analysis is restricted to the conceptual level. He leaves unanswered major questions about origins.

> In justice as fairness the original position of equality corresponds to the state of nature in the traditional theory of the social contract. This original position is not, of course, thought of as an actual historical state of culture. It is understood as a purely hypothetical situation characterized so as to lead to a certain conception of justice. (Rawls, 1971, p. 12)

This is all very well. But "purely hypothetical situations" are hardly satisfying. Interestingly, as hinted above, Rawls himself suggests that biology might be important.

> In arguing for the greater stability of the principles of justice I have assumed that certain psychological laws are true, or approximately so. I shall not pursue the question of stability beyond this point. We may note however that one might ask how it is that human beings have acquired a nature described by these psychological principles. The theory of evolution would suggest that it is the outcome of natural selection; the capacity for a sense of justice and the moral feelings is an adaptation of mankind to its place in nature. (Rawls, 1971, pp. 502–503)

This is precisely the evolutionist's approach. There is no need to suppose hypothetical contracts. Natural selection made us as we are.

4. *Metaethics*

Many traditional philosophers will feel able to go this far with the evolutionist. But now serious differences will begin to appear about the foundations of such ethics, that is, about metaethics. The argument will run like this:[1]

The evolution of ethics has nothing to do with the status of ethics. I may be kind to others because my biology tells me to be kind to others, and because those proto-humans who were not kind to others failed to survive and reproduce. But is it right that I be kind to others? Do I really, objectively, truly have moral obligations? To suppose that the story of origins tells of truth or falsity is to confuse causes with reasons. In a Spencerian fashion, it is to jumble the way things came about with the way things really are. Or ought to be. Since Rawls has been quoted as an authority, let us recall what he says at the end of his speculations on the evolution of morality: "These remarks are not intended as justifying reasons for the contract view" (Rawls, 1971, p. 504).

This is a powerful response, but today's evolutionary ethicist argues that it misses entirely the full force of what biology tells us. It is indeed true that you cannot deduce moral claims from factual claims (about origins). However, using factual claims about origins, you can give moral claims the only foundational explanation that they might possibly have. In particular, the evolutionist argues that, thanks to our science, we see that claims such as, "You ought to maximize personal liberty" are no more than subjective expressions, impressed upon our thinking because of their adaptive value. In other words, we see that morality has no objective foundation. It is just an illusion, fobbed off on us to promote "altruism."

This is a strong claim, so let us understand it fully. The evolutionist is no longer attempting to derive morality from factual foundations. The claim now is that there are no foundations of any sort from which to derive morality—whether these foundations are evolution, God's will, or whatever. Since, clearly, ethics is not nonexistent, the evolutionist locates our moral feelings simply in the subjective nature of human psychology. At this level, morality has no more (and no less) status than that of the terror we feel at the unknown—another emotion which undoubtedly has good biological adaptive value.

Consider an analogy. During the First World War, many bereaved parents turned to spiritualism for solace. Down the ouija board would come the messages: "It's alright, Mum. I've gone to a far better place. I'm just waiting for you and Dad." I take it that these were not in fact the words of the late Private Higgins, speaking from beyond. Rather, they were illusory—a function of human psychology, as persons projected their wishes.[2]

The moral to be drawn from this little story is that we do not need any further justificatory foundation for "It's alright, Mum" than that just given. At this point, we do not need a reasoned underpinning to the words of reassurance. ("Why is it alright?" "Because I'm sitting on a cloud, dressed in a bed-sheet, playing a harp.") What we need is a causal explanation of why the bereaved "heard" what they did. The evolutionist's case is that something similar is very true of ethics. Ultimately, there is no reasoned justification for ethics, in the sense of foundations to which one can appeal in reasoned argument. All one can offer is a causal argument to show why we hold ethical beliefs. But, once such an argument is offered, we can see that this is all that is needed.

In a sense, therefore, the evolutionist's case is that ethics is a collective illusion of the human race, fashioned and maintained by natural selection in order to promote individual reproduction. Yet, more must be said than this. Obviously, "Stamping on small children is wrong," is not really illusory like "It's alright, Mum. I'm okay!" However, we can easily show why the analogy breaks down at this point. Morality is a shared belief (or set of beliefs) of the human race, unlike the messages down the ouija board. Thus, we can distinguish between "Love little children," which is certainly not what we would normally call "illusory," and "Be kind to cabbages on Fridays," which certainly is what we would normally call "illusory." We all (or nearly all) believe the former, but not the latter.

Perhaps we can more accurately express the evolutionist's thesis by drawing back from a flat assertion that ethics is illusory. What is really important to the evolutionist's case is the claim that ethics is illusory inasmuch as it persuades us that it has an objective reference. This is the crux of the biological position. Once it is grasped, everything falls into place.

This concession about the illusory status of ethics in no way weakens the evolutionist's case. Far from it! If you think about it, you will see that the very essence of an ethical claim, such as "Love little children," is that, whatever its true status may be, we think it binding upon us, because we think it has an objective status. "Love little children" is not like "My favorite vegetable is spinach." The latter is just a matter of subjective preference. If you do not like spinach, then nothing ensues. But we do not take the former, the moral claim, to be just a matter of preference. It is regarded as objectively binding upon us—whether we take the ultimate source of this objectivity to be God's will, or (if we are Platonists) intuited relations between the Forms, or (like G. E. Moore) apprehension of nonnatural properties, or whatever.

The evolutionist's claim, consequently, is that morality is subjective—it is all a question of human feelings or sentiments—but he or she admits that we "objectify" morality, to use an ugly but descriptive term. We think morality has objective reference, even though it does not. And because of this, a causal analysis of the type offered by the evolutionist is appropriate and adequate, whereas a justification of moral claims in terms of reasoned foundations is neither needed nor appropriate. Further, completing the case, the evolutionist points out that there are good biological reasons why it is part of our nature to objectify morality. If we did not regard it as binding, we would ignore it. It is precisely because we think that morality is more than mere subjective desires that we are led to obey it.[3]

5. *Objections*

A host of questions will be raised. I will concentrate on two of the more important. First, let us turn to a substantive question. Many queries at this level will be based on misunderstandings of the evolutionist's position. For instance, although the evolutionist is subjectivist about ethics, this does not in any sense imply that he or she is a relativist—especially not a cultural relativist. The whole point about the evolutionary approach to ethics is that morality does not work unless we are all in the game (though we can perhaps tolerate a few cheaters, so-called "criminals" or "sociopaths," who must also be suppressed lest they undermine the game). Moreover, we have to believe in morality, otherwise it will not work. Hence, the evolutionist looks for shared moral insights, and cultural variations are dismissed as mere fluctuations due to contingent impinging factors.

Analogously, there is no question of simply breaking from morality, if we so wish. Even though we have insight into our biological nature, it is still our biological nature. We can certainly do immoral things. We do them all the time. But, a policy of persistently and consistently breaking the rules can only lead to internal tensions. Plato had a good point in the *Republic* when he argued that only the truly good man is the truly happy man.

A much more significant question concerns reciprocation. No one should be misled into thinking that the evolutionist proclaims the virtues (moral or otherwise) of selfishness, or that the evolutionist's position implies that, as a matter of contingent fact, we are totally selfish. Yes, sociobologists find, and admit, that human beings have a tendency toward selfishness; but you did not need an evolutionist to tell you that. What is surprising is that we are not totally selfish. Humans have genuinely altruistic feelings toward their fellows. The fact that (according to the evolutionist) we are brought to (literal) altruism, by our genes acting in our biological self-interests, says nothing against the genuineness of our feelings.

Nevertheless, while this is indeed all true, a nagging doubt remains. Let us look for a moment at the actual causal models proposed by sociobiologists in order to explain the evolution of altruism. First, it is suggested that kin selection is important. Relatives share copies of the same genes. Hence, to the degree to which my relative reproduces, I myself reproduce vicariously. Therefore, help given to relatives leading to survival and reproduction rebounds to my own benefit. Second, there is reciprocal altruism. Simply, if I help you (even though you are no relative), then you are more likely to help me—and conversely. We both gain together, whereas apart we both lose.[4]

Now, surely, with both of these mechanisms, the possibility of genuine altruism seems precluded. With kin selection, the rewards come through your relatives' reproduction, so there is no need for crude, overt returns. But, would not mere nonmoral love do all that is needed? I love my children and I help them, not because it is right, but because I love them. As Immanuel Kant ([1785]1959) rightly points out, unless you are actually heeding the call of duty, there is no moral credit. A mother happily suckling her baby is not performing a moral act.

In the case of reciprocal altruism, the problems for the evolutionary ethicist are even more obvious. You do something in hope of return. This is not genuine altruism, but a straight bargain. There is nothing immoral in such a transaction. If I pay cash for a kilo of potatoes, there is no wrongdoing. But, there is nothing moral in such a transaction, either. Morality means going out on a limb, because it is right to do so. Morality vanishes if you hope for payment.

The evolutionist has answers to these lines of criticism—answers that strengthen the overall position. First, it is indeed true that much we do for our family stems from love, without thought of duty. But, only the childless would think moral obligations never enter into intrafamilial relations. Time and again we have to drive ourselves on, and we do it because it is right. Without the concepts of right and wrong, we would be much less successful parents (uncles, aunts, relatives) than we are. Humans require so much child care that they make the case for a biological backing to morality particularly compelling. If parental duties were left to feeling of kindliness, the system would break down.[5]

Second, reciprocal altruism would fail if there were no returns—or ways of enforcing returns. But it is not necessary to suppose that such reciprocation requires a direct demand of returns for favors granted, one-on-one, or person-by-person. Morality is like a group insurance policy, rather than a person-to-person transaction. I help you but do not necessarily expect you personally to help me. Rather, my help is thrown into the general pool, as it were, and then I am free to draw on help as needed.

Furthermore, enforcement of the system comes about through morality itself! I help you, and I can demand help in return, not because I have helped you or even because I want help, but because it is right that you help me. Reciprocation is kept in place by moral obligations. If you cease to play fair, then before long I and others will chastise you or take you out of the moral sphere. We do this because you are too "sick" to recognize the right way of doing things, despite the fact that we may still like you. Morality demands that we give freely, but it does not expect us to make suckers of ourselves. (What about Jesus' demand that we forgive seven times seventy times? The moral person responds that forgiveness is one thing, but that complacently letting a bad act occur 490 times borders on the criminally irresponsible. We ought to put a stop to such an appalling state of affairs.[6])

Thus far there is little in the evolutionist's approach to normative ethics, properly understood, which should spur controversy. But, let me conclude this section by pointing to one implication that will certainly cause debate. Many moralists argue that we have an equal obligation to all human beings, indifferent as to relationship acquaintance, nationality, or whatever (Singer, 1972). In principle, my obligations to some unknown child in (say) Ethiopia are no less than to my own son. Nevertheless, although many (most?) would pay lip-service to some such view as this, my suspicion is that, sincerely meant, this doctrine makes the evolutionist decidedly queasy. Biologically, our major concern has to be toward our own kin, and then to those in at least some sort of relationship to us (not necessarily a blood relationship), and only finally to complete strangers. And, feelings of moral obligation have to mirror biology.

I speak tentatively now. You could argue that biology gives us an equal sense of obligation toward all, and that this sense is then filtered across strong (nonmoral) feelings of warmth toward our own children, followed by diminishing sentiments toward nonrelatives, ending with a natural air of suspicion and indifference toward strangers. But, my hunch is that the care we must bestow on our children is too vital to be left to chance, and therefore we expect to find (what we do in fact find) that our senses of obligation vary over the communities within which we operate. Therefore, whatever we may sometimes say, truly we have a stronger feeling of moral obligation toward some people than toward others.

It is perhaps a little odd to speak this hesitantly about our own feelings, including moral feelings. You might think that one should be able to introspect and speak definitively. However, matters are not always quite this simple, particularly when (as now) we are faced with a case where our technology has outstripped our biology (and consequent morality). A hundred years ago, it would have made little sense to talk of moral obligations to Ethiopians. Now, we know about Ethiopians and, at least at some level, we can do something for them. But, what should we do for them? Within the limits of our abilities, should we do as much for each Ethiopian child as for each of our own children? I suspect that most people would say not. I hasten to add that no evolutionist says we have no obligations to the world's starving poor. The question is whether we have a moral obligation to beggar our families and to send all to Oxfam.

Let me at least note that, over this matter of varying obligations, the evolutionist takes no more stringent a line than does Rawls. Explicitly, Rawls treats close kin as a case meriting special attention, and, as he himself admits, it is far from obvious that his theory readily embraces relations with the Third World (Rawls, 1980). It is not intuitively true that, even hypothetically, we were in an original position with the people of Africa, India, or China. Hence, although the evolutionist certainly does not want to hide behind the cloth of the more conventionally moral philosopher, he or she can take comfort from the fact that he or she is in good company. There is even Bible verse about this: "If any one does not provide for his relatives, and especially for his own family, he has disowned the faith and is worse than an unbeliever" (1 Timothy 5.8). And it is an old proverb that "Charity begins at home."

6. *Objectivity*

We turn now to foundational, metaethical worries. The central claim of the evolutionist is that ethics is subjective, a matter of feelings or sentiment, without genuine objective referent. What distinguishes ethics from other feelings is our belief that ethics is objectively based, and it is because we think this that ethics works. The most obvious and important objection to all of this is that the evolutionist has hardly yet really eliminated the supposedly objective foundation of morality. Of course, ethics is, in some way, objective. How could it not be? It is a system of beliefs held by humans. But, this does not in itself deny that there is something more. Consider, analogously, the case of perception. I see the apple. My sensations are subjective, and my organs of vision (eyes) came through the evolutionary process, for excellent biological reasons. Yet, no one would deny that the apple is independently, objectively real. Could not the same be true of ethics? Ultimately, ethics resides objectively in God's will, or some such thing.[7]

Let us grant the perception case, although, parenthetically, I suspect the evolutionist might well have some questions about the existence of a real world, beyond the knowing subject. The analogy with ethics still breaks down. Imagine two worlds, identical except that one has an objective ethics (whatever that might mean) and one does not. Perhaps, in one world, God wants us to look after the sick, and in the other God could not care less. The evolutionist argues that, in both situations, we would have evolved in such a way as to think that, morally, we ought to care for the sick. To suppose otherwise, to suppose that only the world of objective ethics has us caring about the sick, is to suppose that there are extra-scientific forces at work, directing and guiding the course of evolution. And this supposition is an anathema to the modern biologist (Ruse, 1982).

In other words, in the light of what we know of evolutionary processes, the objective foundation has to be judged redundant. But, if anything is a contradiction in terms, it is a redundant objective morality. "The only reason for loving your neighbor is that God wants this, but you will think you ought to love your neighbor whether or not God wants it." In fact, if you take seriously the notion that humans are the product of natural selection, the situation is even worse than this. We are what we are because of contingent circumstances, not because we necessarily had to be as we are. Suppose, instead of evolving from savannah-living primates (which we did), we had come from cave dwellers. Our nature and our morality might have been very different. Or, take the termites (to go to an extreme example from a human perspective). They have to eat each others feces, because they lose certain parasites, vital for digestion, when they molt. Had humans come along a similar trail, our highest ethical imperatives would have been very strange indeed.

What all this means is that, whatever objective morality may truly dictate, we might have evolved in such a way as to miss completely its real essence. We might have developed so that we think we should hate our neighbors, when really we should love them. Worse than this, perhaps we really should be hating our neighbors, even though we think we should love them! Clearly, this possibility reduces objectivity in ethics to a mass of paradox.

But does it? Let us grant that the evolutionist has a good case against the person who would argue that the foundations of morality lie in sources external to us humans, whether these sources are God's will, or the relations of Platonic Forms, or nonnatural properties, or whatever. However, there is at least one well-known attempt to achieve objectivity (of a kind), without the assumption of externality. I refer to the theories of Immanuel Kant ([1788]1949, [1785]1959). He argued that the supreme principle of morality, the so-called Categorical Imperative, has a necessity that transcends the contingency of human desires. It is synthetic *a priori*, by which Kant meant that morality is a condition that comes into play, necessarily, when rational beings interact. He argued that a disregard of morality leads to "contradictions," that is, to a breakdown in social functioning. Thus, we see that morality is not just subjective whim, but has its being in the very essence of rational interaction. To counter an example offered previously, we could not have evolved as pure haters, because such beings simply could not interact socially.

Since, more than once in my arguments, the evolutionist has invoked the ideas of John Rawls, a critic might reasonably point out that (having left matters dangling in *A Theory of Justice*), more recently Rawls has tried explicitly to put morality on a Kantian foundation. At a general level, he writes:

> What justifies a conception of justice is not its being true to an order antecedent to and given to us, but its congruence with our deeper understanding of ourselves and our aspirations, and our realization that, given our history and the traditions embedded in our public life, it is the most reasonable doctrine for us. (Rawls, 1980, p. 519)

Further explaining, Rawls claims:

> Kantian doctrine interprets the notion of objectivity in terms of a suitably constructed social point of view that is authoritative with respect to all individual and associational points of view. This rendering of objectivity implies that, rather than think of the principles of justice as true, it is better that they are the principles most reasonable for us, given our conception of persons as free and equal, and fully cooperating members of a democratic society. (Rawls, 1980, p. 554)

Thus, in some way, we try to show both that morality is reasonable and that it is more than a matter of mere desire or taste, like a preference for vegetables.

Responding to the Kantian/Rawlsian position, sometimes called a "constructivist" position, the evolutionist will want to make two points.

First, there is much in the position with which he or she heartily sympathizes! Both constructivist and evolutionist agree that morality must not be sought outside human beings, and yet both agree that there is more to morality than mere feelings. Additionally, both try to make their case by pointing out that morality is the most sensible strategy for an individual to pursue. Being nice pays dividends—although, as both constructivist and evolutionist point out, one behaves morally for good reasons, not because one is consciously aware of the benefits.

Second, for all of the sympathy, the evolutionist will feel compelled to pull back from the full conclusions of the constructivist position. The evolutionist argues that morality (as we know it) is the most sensible policy, as we humans are today. However, he or she draws back from the constructivist claim that (human-type) morality must be the optimal strategy for any rational being. What about our termite-humans, for instance? They might be perfectly rational.

Possibly, the response will be that the termite-humans' sense of obligation to eat rather strange foodstuffs is covered by a prohibition against suicide, which Kant certainly thinks follows from the Categorical Imperative. Hence, the constructivist admits that one's distinctive (in our case, human) nature gives one's actual morality a correspondingly distinctive appearance, but then argues that underlying the differences is a shared morality. The principle is the same as when everyone (including the evolutionist) explains differences in cultural norms as due to special circumstances, not to diverse ultimate moral commitments (Taylor, 1958).

Yet, the evolutionist continues the challenge. If the constructivist argues that the only thing that counts is rational beings working together, and that their contingent nature is irrelevant, then it is difficult to see why morality necessarily emerges at all. Suppose that we had evolved into totally rational beings, with the super-brains that I mentioned earlier, and that we calculated chances, risks, and benefits, at all times. We would be neither moral nor immoral, feeling no urges of obligation at all.

Obviously, we are not like this. Apparently, therefore, we must take account of an evolved being's contingent nature—no matter how rational it may be—in order to get some kind of morality. But this is the thin end of the wedge for moralities other than human morality. Think, for instance, how we might patch up the society of pure haters, so a kind of morality could emerge—and this a kind quite different from ours. Suppose that it is part of our nature to hate others, and that we think we have an obligation to hate others. A Kantian "contradiction" (that is, breakdown in sociality) might still be avoided, and cooperation achieved, because we know that others hate us and so we feel we had better work warily together to avoid their wrath. If this sounds far-fetched, consider how today's supposed superpowers function—or used to function! Everything would be perfectly rational and could work (after a fashion). There would be a cooperative standoff born of mutual hate. Yet, there would be little that we humans would recognize as "moral" in any of this.

You can still point out, if you wish, that such a society of pure haters would end up with rules much akin to those that the constructivist endorses, about liberty and so forth. But, these rules would not be moral in any sense. They would be, explicitly, rules of expediency, of self-interest. I give you liberty, not because I care for you, or respect you, or think I ought to treat you as a worthwhile individual. I hate your guts! And, I think I ought to hate you. I give you liberty simply because it is in my consciously thought-out interests to do so. This may be a sensible policy. It is not a moral policy.

The evolutionist concludes, against the constructivist, that our morality is a function of our actual human nature, and it cannot be divorced from the contingencies of our evolution. Morality, as we know it, cannot have the necessity or objectivity sought by the Kantian and Rawlsian.

7. *Biology and Ethics*

Our biology is working hard to make the evolutionist's position seem implausible. We are convinced that morality really is objective, in some way. However, if we take modern biology seriously, we come to see how we are children of our past. We learn what the true situation really is. Evolution and ethics are at last united in a profitable symbiosis, and this is done without committing all of the fallacies of the last century. We have succeeded in a unified theory combining biology, ethics, and the origins of human life, a theory that can and ought to govern our contemporary and future practice.

References

Betz, D. 1985. *Essays on the Sermon on the Mount*. Philadelphia: Fortress.

Darwin, Charles. 1859. *On the Origin of Species*. London: John Murray.

Dawkins, Richard. 1976. *The Selfish Gene.* Oxford: Oxford University Press.

Flew, A. G. N. 1967. *Evolutionary Ethics.* London: Macmillan.

Freud, Sigmund. [1929–30]1961. *Civilization and Its Discontents.* In J. Strachey ed., *Complete Psychological Works of Sigmund Freud* (vol. 21, pp. 64–145). London: Hogarth Press.

Hume, David. [1739]1978. *A Treatise of Human Nature.* Oxford: Clarendon Press.

Huxley, T. H. 1901. *Evolution and Ethics.* London: Macmillan.

Kant, Immanuel. [1788]1949. *Critique of Practical Reason.* Chicago: University of Chicago Press.

Kant, Immanuel. [1875]1959. *Foundations of the Metaphysics of Morals.* Indianapolis: Bobbs-Merrill.

Lovejoy, O. 1981. "The origin of man." *Science* 211: 341–350.

Lumsden, Charles J., and Edward O. Wilson. 1981. *Genes, Mind and Culture: The Coevolutionary Process.* Cambridge, MA: Harvard University Press.

Lumsden, Charles J., and Edward O. Wilson. 1983. *Promethean Fire.* Cambridge, MA: Harvard University Press.

Mackie, J. L. 1977. *Ethics: Inventing Right and Wrong.* Harmondsworth, Middlesex: Penguin.

Mackie, J. L. 1978. "The law of the jungle." *Philosophy* 53: 553–73.

Moore, G. E. 1903. *Principia Ethica.* Cambridge: Cambridge University Press.

Murphy, Jeffrie G. 1982. *Evolution, Morality, and the Meaning of Life.* Totowa, NJ: Rowman and Littlefield.

Nozick, Robert. 1981. *Philosophical Explanations.* Cambridge, MA: Harvard University Press.

Quinton, A. 1966. "Ethics and the theory of evolution." In I. T. Ramsey, ed., *Biology and Personality.* Oxford: Blackwell.

Raphael, D. D. 1958. "Darwinism and ethics." In S. A. Barnett, ed., *A Century of Darwin.* London: Heinemann.

Rawls, John. 1971. *A Theory of Justice.* Cambridge, MA: Harvard University Press.

Rawls, John. 1980. "Kantian constructivism in moral theory." *Journal of Philosophy*, 77, 515–572.

Ruse, Michael. 1979a. *The Darwinian Revolution: Science Red in Tooth and Claw.* Chicago: University of Chicago Press.

Ruse, Michael. 1979b. *Sociobiology: Sense or Nonsense?* Dordrecht: Reidel.

Ruse, Michael. 1982. *Darwinism Defended: A Guide to the Evolution Controversies.* Reading, MA: Addison-Wesley.

Ruse, Michael. 1985. *Taking Darwin Seriously: A Naturalistic Approach to Philosophy.* Oxford: Blackwell.

Ruse, Michael, and Edward O. Wilson. 1986. "Darwinism as applied science." *Philosophy* 61:173–192.

Russett, C.E. 1976. *Darwin in America.* San Francisco: W. H. Freeman.

Singer, Peter. 1972. "Famine, affluence, and morality." *Philosophy and Public Affairs* 1: 229–243.

Spencer, Herbert. 1852. "A theory of population, deduced from the general law of animal fertility." *Westminster Review* 1: 468–501.

Spencer, Herbert. 1857. "Progress: its law and cause." *Westminster Review.* Reprinted in *Essays: Scientific, Political, and Speculative* (vol. 1, pp. 1–60). London: Williams and Norgate.

Taylor, Paul W. 1958. "Social science and ethical relativism." *Journal of Philosophy* 55: 32–44.

Taylor, Paul W. 1978. *Problems of Moral Philosophy.* Belmont, CA: Wadsworth.

Trivers, Robert L. 1971. "The evolution of reciprocal altruism." *Quarterly Review of Biology* 46: 35–57.

Williams, George C. 1966. *Adaptation and Natural Selection.* Princeton: Princeton University Press.

Wilson, Edward O. 1971. *The Insect Societies.* Cambridge, MA: Belknap Press.

Wilson, Edward O. 1975. *Sociobiology: The New Synthesis.* Cambridge, MA: Harvard University Press.

Wilson, Edward O. 1978. *On Human Nature.* Cambridge, MA: Harvard University Press.

Endnotes

1. Versions of this argument occur in Raphael (1958), Quinton (1966), Singer (1972), and—I blush to say it—Ruse (1979b).

2. We can, I think, discount the view that all or even most of these experiences are fraud.

3. See Murphy (1982) for more on the argument that a causal explanation might be all that can be offered for ethics, and Mackie (1977) for discussion of "objectification" in ethics.

4. These two mechanisms are discussed in detail in Ruse (1979b). They are related to human behavior in some detail in Wilson (1978).

5. There has been a feedback causal process at work here. Because humans have evolved a moral capacity, child care has become more extended; and, the needs of extensive child care, in turn, set up pressures toward increased moral awareness.

6. This criticism assumes that the Christian is obligated to forgive endlessly, without response. Modern scholarship suggests that this is far from Jesus' true message. See Betz (1985) for more on this point, and Mackie (1978) for more on the sociobiologically inspired criticism that Christianity makes unreasonable demands on us. This latter line of argument obviously parallels that of Freud in *Civilization and its Discontents* (1961).

7. Nozick (1981) pursues a line of argument akin to this.

5

The Difference of Being Human

Ethical Behavior as an Evolutionary Byproduct

FRANCISCO J. AYALA

■ ■ ■ *Editor's Introduction*

Having heard Ruse, a philosopher of biology, claim that ethics is literally a survival tool and nothing more, Francisco J. Ayala, a geneticist, demurs. Not so. Ethics is a only a byproduct of selection for intelligence in the hominoid line. With increasing intelligence reaching the large brains of *Homo sapiens* there arises: "(a) the ability to anticipate the consequences of one's own actions; (b) the ability to make value judgments; and (c) the ability to choose between alternative courses of action" (p. 118). These features contribute to survival; if I can evaluate, choose, anticipate consequences, and then act, I shall more likely survive than a competitor who cannot do these things, or does them less well.

But, curiously, it just so happens that these three gifts of general intelligence are exactly "the three necessary, and jointly sufficient, conditions for ethical behavior" (p. 118). Or perhaps this is only curious biologically. It is not curious logically; rather it is logically inevitable. Exactly the same factors that are required for general intelligence are required for conscience. Nevertheless, Ayala holds that intelligence is the target and conscience the byproduct. He does not so conclude on logical grounds, since the required conditions for one are the same as the required conditions for the other. He identifies conscience as byproduct on empirical, biological grounds. There is no cause to think that conscience would contribute to survival; there is much cause to think that other activities of intelligence do.

Ayala easily finds evidence that, from the earliest hominids to the present, tool-making is adaptive, and that, in modern cultures, science is adaptive. But Ayala finds no evidence that ethical behavior is adaptive. Contrary to Ruse who thinks ethics is quite adaptive, Ayala thinks that it is only byproduct. Afterward, having disconnected ethics from survival, Ayala is free to hold that the normative content of ethics is culturally based, not biologically driven. "Moral norms are products of cultural evolution, not of biological evolution" (p. 118).

Readers may see a dilemma arising. On the one hand, it is difficult to say, with Ruse, that ethics is nothing but an adaptation for survival, and we may welcome Ayala's prospect of an evaluation of right and wrong within culture, with the norms freed from biology. On the other hand, we may be reluctant to conclude that ethics makes no contribution at all to survival. We begin to wonder whether ethics is just a byproduct. Maybe ethics is adaptive, in the sense that groups that cooperate do well. The ethical systems sometimes are urged because they bring "the greatest good for the greatest number," which presumably includes prosperity and success in child rearing. Ayala notices that no ethical system can really gainsay human nature, and part of human nature is the need to reproduce and rear a next generation.

Biologically speaking, any ethic will do, so long as it fits human nature generically and produces a next generation. From such a set of ethical systems, we may and ought further to choose the better from the worse ones using cultural norms. Human biology is not overridden; it cannot be "counteracted" (p. 119); but neither can we use biology to select from among the permitted options. It would commit, Ayala thinks, the naturalistic fallacy were we to move from biology to ethics. Meanwhile, given the logical accompaniment of ethics with advanced intelligence, such a move is thrust upon us by nature. In that sense, nature accidentally but inevitably gives humans an assignment without criteria by which it can be evaluated. We are inevitably moral, but it is a fallacy to think that nature gives us any guidance. "Biology is insufficient for determining which moral codes are, or should be, accepted" (p. 134).

One response will be to welcome this freedom and responsibility. The content of our ethics is not genetically determined but open to our choice. Another response, however, recalling Whitehead's caution, which will increasingly become a theme in this volume, is to begin to worry whether ethics is not some kind of a possibility floating in from nowhere. Perhaps it is only biologically speaking that ethical beliefs are byproducts; in a more comprehensive world view, with a metaphysics, we may have to say more—as Gilkey and Birch will soon maintain. Meanwhile, the origin of ethics is an interesting story in serendipity and epiphenomena.

Has ethics a significant survival value, in helping humans to cooperate so as to leave more offspring in the next generation? Compare the significant differences here between Ruse and Ayala. Does the advanced intelligence of humans both logically and biologically result in a moral conscience? What are the connections between the power to deliberate and to choose and the capacity to be responsible? Is it an adequate explanation of ethics to discover that it is a byproduct? If ethics must be consistent with human nature, does this mean that our biology to some extent controls our ethics? If human evolution produces an intelligence with the capacity for ethics but fails to supply norms, where do these norms come from? How are such norms evaluated from culture to culture? What are the most reasonable criteria for ethical norms?

Born in Spain, Francisco J. Ayala is Professor of Biological Sciences and also Professor of Philosophy at the University of California at Irvine. He is the author of *Molecular Evolution* (1976), *Modern Genetics* (2nd ed., 1984), *Population and Evolutionary Genetics: A Primer* (1982), *Evolving: The Theory and Processes of Organic Evolution* (1979), *Evolution* (1977), and *Studies in the Philosophy of Biology* (1974), in addition to more than 500 articles. He is co-editor, with Michael Ruse, of the journal *Biology and Philosophy*. He is a member of the National Academy of Sciences and a fellow of the American Association for the Advancement of Science. He has been a Guggenheim Fellow and twice a Fulbright Fellow. He served as an expert witness in the 1981 Arkansas trial on the teaching of evolution. He has lectured in France, Spain, Italy, Denmark, Finland, Norway, Greece, Switzerland, Great Britain, Czechoslovakia, Yugoslavia, Russia, Canada, Mexico, Venezuela, Peru, Brazil, China, and Japan. He is currently (1994–1995) president of the American Association for the Advancement of Science.

■ ■ ■ ─────────────────────────────

1. *A Summary of the Argument*

The last common ancestors to humans and apes lived some eight million years ago. The first hominids were the Australopithecines, who appeared four million years ago; they had a bipedal gait and small brains (about 400 cubic centimeters). The hominid brain gradually increased in size, reaching 1,400 cc within the last one hundred thousand years. Simple stone tools were first made by our ancestors two million years ago; more complex and diversified tools only came about one hundred thousand years ago.

Erect posture and large brain are distinctive anatomical features of modern humans. High intelligence, symbolic language, political institutions, technology, religion, and ethics are some of the behavioral traits that distinguish us from other animals. An explanation of human evolution must account for the various anatomical and behavioral traits in terms of natural selection (and other evolutionary processes). One explanatory strategy is to focus on a particular human feature and seek to identify the conditions under which this feature may have been favored by natural selection. Such a strategy may lead to erroneous conclusions as a consequence of the fallacy of selective attention: some traits may have come about not because they are themselves adaptive, but rather because they are associated with other traits that are favored by natural selection. Geneticists have long recognized the phenomenon of "pleiotropy," the expression of one particular gene in several different organs or anatomical traits. It follows that a gene that becomes changed owing to its effects on a selected trait will result in the modification of other traits as well.

Literature, art, science, technology, and other behavioral features may have come about not because they were adaptively favored in human evolution, but because they are expressions of the high intellectual abilities present in modern humans: what was favored by natural selection (its "target") was an increase in intellectual ability rather than each one of those particular activities. I argue here that ethical behavior (the proclivity to judge human actions as either good or evil) has evolved as a distinctive trait of human behavior not because it was adaptive in itself, but rather as a pleiotropic consequence of the high intelligence characteristic of humans.

I first point out that the question of whether ethical behavior is biologically determined may refer either to the *capacity* for ethics (i.e., the proclivity to judge human actions as either right or wrong) or to the moral *norms* accepted by human beings for guiding their actions. My theses are: (1) that the capacity for ethics is a necessary attribute of human nature; and (2) that moral norms are products of cultural evolution, not of biological evolution.

Humans exhibit ethical behavior by nature because their biological makeup determines the presence of the three necessary, and jointly sufficient, conditions for ethical behavior: (a) the ability to anticipate the consequences of one's own actions; (b) the ability to make value judgments; and (c) the ability to choose between alternative courses of action. Ethical behavior came about in evolution not because it is adaptive in itself, but as a necessary consequence of man's eminent intellectual abilities, which are an attribute directly promoted by natural selection.

Since Darwin's time, there have been evolutionists proposing that the norms of morality are derived from biological evolution. Some sociobiologists have recently developed a quite subtle version of that proposal. The sociobiologists' argument is that human ethical norms are sociocultural correlates of behaviors fostered by biological evolution. I argue that such proposals are misguided and do not escape the naturalistic fallacy. Perhaps it is true that both natural selection and moral norms sometimes coincide on the same behavior. The two are consistent. But this isomorphism between the behaviors promoted by natural selection and those sanctioned by moral norms exists only with respect to the consequences of the behaviors; the underlying causations are completely disparate.

2. *Ethics and Language: A Parallel Distinction*

I just noted that the question of whether ethical behavior is biologically determined may refer to either of the following issues: (1) Is the capacity for ethics—the proclivity to judge human actions as either right or wrong—determined by the biological nature of human beings? (2) Are the systems or codes of ethical norms accepted by human beings biologically determined? A similar distinction can be made with respect to language. The issue of whether the capacity for symbolic language is determined by our biological nature is different from the question of whether the particular language we speak (English, Spanish, or Japanese) is biologically necessary.

The first question is more fundamental; it asks whether or not the biological nature of *Homo sapiens* is such that humans are necessarily inclined to make moral judgments and to accept ethical values, to identify certain actions as either right or wrong. Affirmative answers to this first question do not necessarily determine what the answer to the second question should be. Independently of whether or not humans are necessarily ethical, it remains to be determined whether particular moral prescriptions are in fact determined by our biological nature, or whether they are chosen by society, or by individuals. Even if we were to conclude that people cannot avoid having moral standards of conduct, it might be that the choice of the particular standards used for judgment would be arbitrary. Or that it depended on some other, nonbiological criteria. The need for having moral values does not necessarily tell us what these moral values should be, just as the capacity for language does not determine which language we shall speak.

The thesis that I propose is that humans are ethical beings by their biological nature. Humans evaluate their behavior as either right or wrong, moral or immoral, as a consequence of their eminent intellectual capacities which include self-awareness and abstract thinking. These intellectual capacities are products of the evolutionary process, but they are distinctively human. Thus, I maintain that ethical behavior is not causally related to the social behavior of animals, including kin, and reciprocal "altruism."

A second thesis that I put forward is that the moral norms according to which we evaluate particular actions as morally either good or bad (as well as the grounds that may be used to justify the moral norms) are products of cultural evolution, not of biological evolution. The norms of morality belong, in this respect, to the same category of phenomena as the languages spoken by different peoples, their political and religious institutions, and the arts, sciences, and technology. The moral codes, like these other products of human culture, are often consistent with the biological predispositions of the human species, dispositions we may to some extent share with other animals. But this consistency between ethical norms and biological tendencies is not necessary or universal: it does not apply to all ethical norms in a given society, much less in all human societies.

Moral codes, like any other dimensions of cultural systems, depend on the existence of human biological nature and must be consistent with it in the sense that they could not counteract it without promoting their own demise. Moreover, the acceptance and persistence of moral norms is facilitated whenever they are consistent with biologically conditioned human behaviors. But the moral norms are independent of such behaviors in the sense that some norms may not favor, and may hinder, the survival and reproduction of the individual and its genes, which are the targets of biological evolution. Discrepancies between accepted moral rules and biological survival are, however, necessarily limited in scope or would otherwise lead to the extinction of the groups accepting such discrepant rules.

3. *The Necessary Conditions for Ethical Behavior*

The question of whether ethical behavior is determined by our biological nature must be answered in the affirmative. By "ethical behavior" I mean here to refer to the urge *to judge* human actions as either good or bad, which need not require actually choosing *good behavior* (i.e., choosing to do what is perceived as good instead of what is perceived as evil). Humans exhibit ethical behavior by nature because their biological constitution determines the presence in them of the three necessary, and jointly sufficient, conditions for ethical behavior. These conditions are (a) the ability to anticipate the consequences of one's own actions; (b) the ability to make value judgments; and (c) the ability to choose between alternative courses of action. I shall briefly examine each of these abilities and show that they exist as a consequence of the eminent intellectual capacity of human beings.

The ability to anticipate the consequences of one's own actions is the most fundamental of the three conditions required for ethical behavior. Only if I can anticipate that pulling the trigger will shoot the bullet, which in turn will strike and kill my enemy, can the action of pulling the trigger be evaluated as nefarious. Pulling a trigger is not in itself a moral action; it becomes so by virtue of its relevant consequences. My action has an ethical dimension only if I anticipate these consequences.

The ability to anticipate the consequences of one's actions is closely related to the ability to establish the connection between means and ends; that is, of seeing a mean precisely as mean, as something that serves a particular end or purpose. This ability to establish the connection between means and their ends requires the ability to anticipate the future and to form mental images of realities not present or not yet in existence.

The ability to establish the connection between means and ends happens to be the fundamental intellectual capacity that has made possible the development of human culture and technology. The evolutionary roots of this capacity may be found in the evolution of the erect position, which transformed the anterior limbs of our ancestors from organs of locomotion into organs of manipulation. The hands thereby gradually became organs adept for the construction and use of objects for hunting and other activities that improved survival and reproduction, that is, they increased the reproductive fitness of their carriers. The construction of tools depends not only on manual dexterity, but in perceiving them precisely as tools, as objects that help to perform certain actions, that is, as means that serve certain ends or purposes: a knife for cutting, an arrow for hunting, an animal skin for protecting the body from the cold. Natural selection promoted the intellectual capacity of our biped ancestors, because increased intelligence facilitated the perception of tools as tools, and therefore their construction and use, with the ensuing amelioration of biological survival and reproduction.

The development of the intellectual abilities of our ancestors took place over two million years or longer, gradually increasing the ability to connect means with their ends and, hence, the possibility of making ever more complex tools serving remote purposes. The ability to anticipate the future, essential for ethical behavior, is therefore closely associated with the development of the ability to construct tools, an ability that has produced the advanced technologies of modern societies and that is largely responsible for the success of humankind as a biological species. From its obscure beginnings in Africa, humankind has spread over the whole earth except the frozen wastes of Antarctica, and has become the most numerous species of mammal. Numbers may not be an unmixed blessing, but they are a measure of biological success.

The second condition for the existence of ethical behavior is the ability to make value judgments, to perceive certain objects or deeds as more desirable than others. Only if I can see the death of my enemy as preferable to his or her survival (or vice versa) can the action leading to his or her demise be thought as moral. If the alternative consequences of an action are neutral with respect to value, the action cannot be characterized as ethical. The ability to make value judgments depends on the capacity for abstraction, that is, on the capacity to perceive actions or objects as members of general classes. This makes it possible to compare objects or actions with one another and to perceive some as more desirable than others. The capacity for abstraction requires an advanced intelligence such as exists in humans and apparently in them alone.

The third condition necessary for ethical behavior is the ability to choose between alternative courses of action. Pulling the trigger can be a moral action only if I have the option not to pull it. A necessary action beyond our control is not a moral action: the circulation of the blood or the process of food digestion are not moral actions.

Whether there is free will is a question much discussed by philosophers, and this is not the appropriate place to review the arguments. I will only advance two considerations that are common-sense evidence of the existence of free will. One is our personal experience, which indicates that the possibility to choose between alternatives is genuine rather than only apparent. The second consideration is that when we confront a given situation that requires action on our part, we are able mentally to explore alternative courses of action, thereby extending the field within which we can exercise our free will. In any case, if there were no free will, there would be no ethical behavior; morality would only be an illusion. The point that I want to make here is, however, that free will is dependent on the existence of a well-developed intelligence, which makes it possible to explore alternative courses of action and to choose one or another in view of the anticipated consequences.

In summary, ethical behavior is an attribute of the biological make-up of humans and, hence, is a product of biological evolution. But I see no evidence that ethical behavior developed because it was adaptive in itself. I find it hard to see how *evaluating* certain actions as either good or evil (not just choosing some actions rather than others, or evaluating them with respect to their practical consequences) would promote the reproductive fitness of the evaluators. Nor do I see how there might be some form of "incipient" ethical behavior that would then be promoted by natural selection. The three necessary conditions for there being ethical behavior are manifestations of advanced intellectual abilities.

It rather seems that the target of natural selection was the development of these advanced intellectual capacities. This development was favored by natural selection because the construction and use of tools improved the strategic position of our biped ancestors. Once bipedalism evolved and tool-using and tool-making became possible, those individuals more effective in these functions had a greater probability of biological success. The biological advantage provided by the design and use of tools persisted long enough so that intellectual abilities continued to increase, eventually yielding the eminent development of intelligence that is characteristic of *Homo sapiens*.

4. *Moral Codes: A Cultural Legacy*

I have answered in the affirmative the first of the two questions I posed: ethical behavior is rooted in the biological make-up of humans. I have also proposed that ethical behavior did not evolve because it was adaptive in itself, but rather as the indirect outcome of the evolution of eminent intellectual abilities. Now I turn to the second question: whether our biological nature also determines which moral norms or ethical codes human beings must obey. My answer is negative. The moral norms according to which we decide whether a particular action is either right or wrong are not specified by biological evolution but by cultural evolution. The premises of our moral judgments are received from religious and other social traditions.

I hasten to add, however, that moral systems, like any other cultural activities, cannot long survive if they run outright contrary to our biology. The norms of morality must be consistent with biological nature, because ethics can only exist in human individuals and in human societies. One might therefore also expect, and it is the case, that accepted norms of morality will often promote behaviors that increase the biological adaptation of those who behave according to them. But this is neither necessary nor indeed always the case.

5. *Theories of Morality*

There are many theories concerned with the rational grounds for morality, such as deductive theories that seek to discover the axioms or fundamental principles that determine what is morally correct on the basis of direct moral intuition; or theories like logical positivism or existentialism, which negate rational foundations for morality, reducing moral principles to emotional decisions or to other irrational grounds. Since the publication of Darwin's theory of evolution by natural selection, philosophers as well as biologists have attempted to find in the evolutionary process the justification for moral norms. The common ground to all such proposals is that evolution is a natural process that achieves goals that are desirable and thereby morally good; indeed, it has produced humans. Proponents of these ideas claim that only the evolutionary goals can give moral value to human action: whether a human deed is morally right depends on whether it directly or indirectly promotes the evolutionary process and its natural objectives.

Herbert Spencer was perhaps the first philosopher seeking to find the grounds of morality in biological evolution. More recent attempts include those of the distinguished evolutionists Julian S. Huxley (1947, 1953) and C. H. Waddington (1960) and of Edward O. Wilson (1975, 1978), founder of sociobiology as an independent discipline engaged in discovering the biological foundations of all social behavior.

In *The Principles of Ethics* (1893), Spencer seeks to replace the Christian faith as the justification for traditional ethical values with a natural foundation. Spencer argues that the theory of organic evolution implies certain ethical principles. Human conduct must be evaluated, like any biological activity whatsoever, according to whether or not it conforms to the life process; therefore, any acceptable moral code must be based on natural selection, the law of struggle for existence. According to Spencer, the most exalted form of conduct is that which leads to a greater duration, extension, and perfection of life; the morality of all human actions must be measured by that standard. Spencer proposes that, although exceptions exist, the general rule is that pleasure goes with that which is biologically useful, whereas pain marks what is biologically harmful. These associations are an outcome of natural selection—thus, while doing what brings them pleasure and avoiding what is painful, organisms improve their chances for survival. With respect to human behavior, we see that we derive pleasure from virtuous behavior and pain from evil actions, associations which indicate that the morality of human actions is also founded on biological nature.

Spencer proposes, as the general rule of human behavior, that anyone should be free to do anything he or she wants, so long as it does not interfere with the similar freedom to which others are entitled. The justification of this rule is found in organic evolution: the success of an individual, plant, or animal depends on its ability to obtain that which it needs. Consequently, Spencer reduces the role of the state to protecting the collective freedom of individuals to do as they please. This *laissez faire* form of government may seem ruthless, because individuals would seek their own welfare without any consideration for others (except for respecting their freedom), but Spencer believes that it is consistent with traditional Christian values. It should be added that although Spencer sets the grounds of morality on biological nature and on nothing else, he admits that certain moral norms go beyond that which is biologically determined; these are rules formulated by society and accepted by tradition.

Social Darwinism, in Spencer's version or in some variant form, was fashionable in European and American circles during the latter part of the nineteenth century and the early years of the twentieth, but it has few or no distinguished intellectual followers at present. Spencer's critics include the evolutionists Julian S. Huxley and C. H. Waddington who, nevertheless, maintain that organic evolution provides grounds for a rational justification of ethical codes. For Huxley, the standard of morality is the contribution that actions make to evolutionary progress, which goes from less to more "advanced" organisms. For Waddington, the morality of actions must be evaluated by their specific contribution to furthering human evolution.

Huxley's and Waddington's views are based on value judgments about what is or is not progressive in evolution. Contrary to Huxley's proposal, there is nothing objective in the evolutionary process itself (i.e., outside human considerations; see Ayala, 1982a) that makes the success of bacteria, which have persisted as such for more than two billion years and in enormous numbers, less "progressive" than that of the vertebrates, even though the latter are more complex. Nor are the insects, of which more than one million species exist, less successful or less progressive from a purely biological perspective than humans or any other mammal species. Waddington fails to demonstrate why the promotion of human biological evolution by itself should be the standard used to measure what is morally good.

6. *The Naturalistic Fallacy*

A more fundamental objection against the theories of Spencer, Huxley, and Waddington—and against any other program seeking the justification of a moral code in biological nature—is that such theories commit the "naturalistic fallacy" (Moore, 1903), which consists in identifying what "is" with what "ought to be." This error was pointed out already by Hume:

> In every system of morality which I have hitherto met with I have always remarked that the author proceeds for some time in the ordinary way of reasoning . . . when of a sudden I am surprised to find, that instead of the usual copulations of propositions, *is* and *is not*, I meet with no proposition that is not connected with an *ought* or *ought not*. This change is imperceptible; but is, however, of the last consequence. For as this *ought* or *ought not* expresses some new relation or affirmation, it is necessary that it should be observed and explained; and at the same time a reason should be given,

for what seems altogether inconceivable, how this new relation can be a deduction from others, which are entirely different from it. (Hume, [1740]1978, p. 469)

The naturalistic fallacy occurs whenever inferences using the terms *ought* or *ought not* are derived from premises that do not include such terms but are rather formulated using the connections *is* or *is not*. An argument cannot be logically valid unless the conclusions contain only terms that are also present in the premises. In order to proceed logically from that which "is" to what "ought to be," it is necessary to include a premise that justifies the transition between the two expressions. But this transition is what is at stake, and one would need a previous premise to justify the validity of the one making the transition, and so on in a regression *ad infinitum*. In other words, from the fact that something *is* the case, it does not follow that it *ought to be* so in the ethical sense; *is* and *ought* belong to disparate logical categories.

Because evolution has proceeded in a particular way, it does not follow that that course is morally right or desirable. The justification of ethical norms on biological evolution or on any other natural process can only be achieved by introducing value judgments, human choices that prefer one over another biological result or process. Biological nature is in itself morally neutral.

It must be noted, moreover, that using natural selection or the course of evolution for determining the morality of human actions may lead to paradoxes. Evolution has produced the smallpox and AIDS viruses. But it would seem unreasonable to accuse the World Health Organization of immorality because of its campaign for total eradication of the smallpox virus, or to label unethical the efforts to control the galloping spread of the AIDS virus. Human hereditary diseases are conditioned by mutations that are natural events in the evolutionary process. But we do not think it immoral to cure or alleviate the pain of persons with such diseases. Natural selection is a natural process that increases the frequency of certain genes and eliminates others, that yields some kinds of organisms rather than others; but it is not a process moral or immoral in itself or in its outcome, in the same way as gravity is not a morally laden force. In order to consider some evolutionary events as morally right and others wrong, we must introduce human values. Moral evaluations cannot be reached simply on the basis that certain events came about by natural processes.

7. *Sociobiology's Proposal: Biology Standing on Its Head*

Edward O. Wilson has urged that "scientists and humanists should consider together the possibility that the time has come for ethics to be removed temporarily from the hands of the philosophers and biologicized" (Wilson, 1975, p. 562). Wilson, like other sociobiologists (Barash, 1977; Wilson, 1978; Alexander, 1979; see also Ruse, 1986, and Ruse, Chapter 4), believes that sociobiology may provide the key for finding a naturalistic basis for ethics. Sociobiology is "the systematic study of the biological basis of all forms of social behavior in all kinds of organisms" (Wilson, 1977) or, in Barash's concise formulation, "the application of evolutionary biology to social behavior" (Barash, 1977, p. ix). Its purpose is "to develop general laws of the evolution and biology of social behavior, which might then be extended in a disinterested manner to the study of human beings" (Wilson, 1977, p. ix). The program is ambitious: to discover the biological basis of human social behavior, starting from the investigation of the social behavior of animals.

The sociobiologist's argument concerning normative ethics is not that the norms of morality can be grounded in biological evolution, but rather that evolution predisposes us to accept certain moral norms, namely those that are consistent with the "objectives" of natural selection. Because of this predisposition, human moral codes sanction patterns of behavior similar to those encountered in the social behavior of animals. The sociobiologists claim that the agreement between moral codes and the goals of natural selection in social groups was discovered when the theories of kin selection and reciprocal altruism were formulated. The commandment to honor one's parents, the incest taboo, the greater blame usually attributed to the wife's adultery than to the husband's, and the ban or restriction of divorce are among the numerous ethical precepts that endorse behaviors that are also endorsed by natural selection, as has been discovered by sociobiology.

The sociobiologists sometimes reiterate their conviction that science and ethics belong to separate logical realms, that one may not infer what is morally right or wrong from a determination of how things are or are not in nature. According to Wilson, "To devise a naturalistic description of human social behavior is to note a set of facts for further investigation, not to pass a value judgment or to deny that a great deal of the behavior can be deliberately changed if individual societies so wish" (1977, p. xiv). Barash puts it this way: "Ethical judgments have no place in the study of human sociobiology or in any other science for that matter. What is biological is not necessarily good" (1977, p. 278). Alexander asks what it is that evolution teaches us about normative ethics or about what we *ought* to do and responds, "Nothing whatsoever" (1979, p. 276).

There is nevertheless some question as to whether the sociobiologists are always consistent with the statements just quoted. Ruse, for instance, defending the sociobiologists, seems to take the *is* that has resulted from natural selection operating to produce preferences within us and identify it with the *ought* of moral life: "The good is simply that which evolution through selection has led us to regard as good" (Ruse, 1984, p. 93). In humans, whatever norms and values have been selected for are ipso facto good.

Wilson writes that "the requirement for an evolutionary approach to ethics is self-evident. It should also be clear that no single set of moral standards can be applied to all human populations, let alone all sex-age classes within each population. To impose a uniform code is therefore to create complex, intractable moral dilemmas" (Wilson, 1975, p. 564). Moral pluralism is, for Wilson, "innate." Biology, then, at the very least helps us to decide that certain moral codes (e.g., all those pretending to be universally applicable) are incompatible with human nature and, therefore, are unacceptable. This is not quite an argument in favor of the biological determinism of ethical norms, but it does approach determinism from the negative side: because the range of valid moral codes is delimited by the claim that some are not compatible with biological nature.

However, Wilson goes further when he writes: "Human behavior—like the deepest capacities for emotional response which derive and guide it—is the circuitous technique by which human genetic material has been and will be kept intact. *Morality has no other demonstrable ultimate function*" (Wilson, 1978, p. 167, emphasis added). How is one to interpret this statement? It is possible that Wilson is simply giving the reason why ethical behavior exists at all. His proposition would, then, be that humans are prompted to evaluate morally their actions as a means to preserve their genes, their biological nature. But this proposition is erroneous. Human beings are by nature ethical beings in the sense I have expounded earlier: they judge morally their actions because of their innate ability for anticipating the consequences of their actions, for formulating value judgments, and for free choice. Human beings exhibit ethical behavior by nature and necessity, rather than because such behavior would help to preserve their genes or serve any other purpose.

Alternatively, Wilson's statement may be read as a justification of human moral codes: the moral codes' function would be to preserve human genes. But this would entail the naturalistic fallacy and, worse yet, would seem to justify a morality that most of us detest. If the preservation of human genes is the purpose that moral norms serve, Spencer's Social Darwinism would seem right; racism or even genocide could be justified as morally correct if they were perceived as the means to preserve those genes thought to be good or desirable and to eliminate those thought to be bad or undesirable. There is no doubt in my mind that Wilson is not intending to justify racism or genocide, but this is one possible interpretation of his words.

8. *Kin Selection and Animal "Altruism"*

I shall now turn to the sociobiologists' proposition that natural selection favors in animals behaviors that are isomorphic with those behaviors that are sanctioned by the moral codes endorsed by most humans.

Evolutionists had for years struggled with finding an explanation for the apparently altruistic behavior of animals. When a predator attacks a herd of zebras, they will attempt to protect the young in the herd, even if they are not their progeny, rather than fleeing. When a prairie dog sights a coyote, it will warn other members of the colony with an alarm call, even though by drawing attention to itself this increases its own risk. There are quite numerous examples of altruistic behaviors of this kind.

Altruism is defined in the dictionary I have at hand (*Webster's New Collegiate*, 2nd ed.) as "regard for, and devotion to, the interests of others." Yet, to speak of animal altruism is not to claim that explicit feelings of devotion or regard are present, but rather that animals act for the welfare of others at their own risk, just as humans are expected to do when behaving altruistically. Regardless of what motivations are or are not present, biologists encounter the problem of how to justify such behaviors in terms of natural selection. Assume, for illustration, that in a certain species there are two alternative forms of a gene ("alleles"), of which one but not the other promotes altruistic behavior. Individuals possessing the altruistic allele will risk their life for the benefit of others, whereas those possessing the nonaltruistic allele will benefit from the altruistic behavior of their peers without risking themselves. Possessors of the altruistic allele will be more likely to die, and the altruistic allele will therefore be eliminated more often than the nonaltruistic allele. Eventually, after some generations, the altruistic allele will be completely replaced by the nonaltruistic one. But then how is it that altruistic behaviors are common in animals without the benefit of ethical motivation?

One major contribution of sociobiology to evolutionary theory is the notion of "inclusive fitness." In order to ascertain the consequences of natural selection, it is necessary to take into account a gene's effects not only on a particular individual but on all individuals possessing that gene. When considering altruistic behavior, we have to weigh not only the risks for the altruistic individual but also the benefits for other possessors of the same allele. Zebras live in herds composed of blood relatives. An allele prompting adults to protect the defenseless young would be favored by natural selection if the benefit (in terms of saving the lives of individuals carrying the same allele) is greater than the cost (due to the increased risk of the protectors). An individual that lacks the altruistic allele and carries instead a nonaltruistic one will not risk its life, but the nonaltruistic allele is partially eradicated with the death of each defenseless relative.

It follows from this line of reasoning that the more closely related the members of a herd or animal group typically are, the more altruistic behavior should be present. This seems generally to be the case. We need not enter here into the details of the quantitative theory developed by sociobiologists in order to appreciate the significance of two examples. The most obvious is parental care. Parents feed and protect their young because each child has half the genes of each parent: the genes are protecting themselves, as it were, when they prompt a parent to care for its young.

The second example is more subtle: the social organization and behavior of certain animals like the honeybee. Worker bees toil building the hive and feeding and caring for the larvae even though they themselves are sterile and only the queen produces progeny. Assume that in some ancestral hive, an allele arises that prompts worker bees to behave as they now do. It would seem that such an allele would not be passed on to the following generation because such worker bees do not reproduce. But such inference is erroneous. Queen bees produce two kinds of eggs: those that remain unfertilized develop into males (which are therefore "haploid," i.e., carry only one set of genes); others are fertilized (hence, are "diploid," i.e., carry two sets of genes) and develop into worker bees and occasionally into a queen. W. D. Hamilton (1964) demonstrated that, with such a reproductive system, daughter queens and their worker sisters share in two-thirds of their genes, whereas daughter queens and their mother share in only one-half of their genes. Hence, the worker bee genes are more effectively propagated by workers caring for their sisters than if they would produce and care for their own daughters. Natural selection can thus explain the existence in social insects of sterile casts, which exhibit a most extreme form of apparently altruistic behavior by dedicating their life to care for the progeny of another individual (the queen).

Sociobiologists point out that many of the moral norms commonly accepted in human societies sanction behaviors also promoted by natural selection (which promotion becomes apparent only when the inclusive fitness of genes is taken into account). Examples of such behaviors are the commandment to honor one's parents, the incest taboo, the greater blame attributed to the wife's than to the husband's adultery, the ban or restriction on divorce, and many others. The sociobiologists' argument is that human ethical norms are sociocultural correlates of behaviors fostered by biological evolution. Ethical norms protect such evolution-determined behaviors as well as being specified by them.

9. *Sociobiology's Fallacy*

I believe, however, that the sociobiologists' argument is misguided and does not escape the naturalistic fallacy (see Ayala, 1980, 1982b, 1987, for more extensive discussion). Consider altruism as an example. Altruism in the biological sense (altruism$_b$) is defined in terms of the population genetic consequences of a certain behavior. Altruism$_b$ is explained by the fact that genes prompting such behavior are actually favored by natural selection (when inclusive fitness is taken into account), even though the fitness of the behaving individual is decreased. But altruism in the moral sense (altruism$_m$) is explained in terms of motivations: a person chooses to risk his or her own life (or incur some kind of "cost") for the benefit of someone else. The isomorphism between altruism$_b$ and altruism$_m$ is only apparent: an individual's chances are improved by the behavior of another individual who incurs a risk or cost. The underlying causations are completely disparate: the ensuing genetic benefits in altruism$_b$; regard for others in altruism$_m$.

The two disparate meanings of altruism are well distinguished by Ruse (1986a, 1986b; Ruse and Wilson, 1986; Ruse, Chapter 4), who has become an ardent proponent of the sociobiologists' thesis concerning the foundations of ethics. Ruse uses quotation marks ("altruism") to signify altruism in the biological sense and to differentiate it from moral altruism, which he writes without the quotation marks. Ruse has articulated perhaps more clearly than anybody else a sociobiological explanation of the evolution of the moral sense; namely that the moral sense—our proclivity to evaluate certain actions as good and others as evil—has evolved so that we behave in ways that improve our fitness, but do not do so in a way that is immediately obvious. The argument runs as follows. Humans tend to be selfish because that usually serves best our fitness. Yet, there are situations where the (inclusive) fitness of our genes is enhanced by cooperation rather than selfishness; examples are cases of "altruistic" behaviors similar to those of adult zebras protecting the young in the herd or to the warning cry of a prairie dog. Natural selection has tricked humans into exhibiting such (biologically) unobvious beneficial behavior by prompting us to evaluate such behavior as morally right, which in turn has necessitated the evolution of the moral sense.

In Ruse's own words, "All such cooperation for personal evolutionary gain is known technically as 'altruism.' I emphasize that this term is rooted in metaphor, even though now it has the just-given biological meaning. There is no implication that evolutionary 'altruism' (working together for biological payoff) is inevitably associated with moral altruism. . . . [Sociobiologists] argue that moral (literal) altruism might be one way in which biological (metaphorical) 'altruism' could be achieved. . . . Literal, moral altruism is a major way in which advantageous biological cooperation is achieved. . . . In order to achieve 'altruism,' we are altruistic! To make us cooperate for our biological ends, evolution has filled us full of thoughts about right and wrong, the need to help our fellows, and so forth" (Ruse, 1986b, pp. 97–99). Ruse thus provides an explicit interpretation of Wilson's statement that I have quoted above: "Human behavior . . . is the circuitous technique by which human genetic material has been and will be kept intact. Morality has no other demonstrable ultimate function."

In my view, this justification of the evolution of the moral sense is misguided. I have argued that we make moral judgments as a consequence of our eminent intellectual abilities, not as an innate way for achieving biological gain. I have also argued that the sociobiologists' position may be interpreted as also requiring that the preferred *norms* of morality be those that achieve biological gain (because that is, in their view, why the moral sense evolved at all). This, in turn, would justify social attitudes that many of us (sociobiologists included) would judge morally obtuse and even heinous. Gilkey (Chapter 7) has elaborated on how even the sociobiologists can and do themselves make moral judgments that transcend any ethics to which they are entitled by their theories. They do not seek their own biological gain at all. According to my account here, those high moral judgments are a consequence of eminent intellectual abilities. Likewise, Sober (Chapter 6) concludes that, when the human brain evolves, there are many side effects, in some of which the human mind attains a selection process of its own, choosing between ideas, deciding between the better or worse, a selection process that "floats free" from the determination of natural selection. My way of saying this is that the norms of ethical systems are to be judged by cultural criteria, even though natural selection requires us all, as an inevitable byproduct of our eminent intellectual abilities, to be moral agents.

The discrepancy between biologically determined behaviors and moral norms—which amounts to a radical flaw in the sociobiologists' argument for a naturalistic foundation for ethics—is enhanced by three additional considerations that I shall briefly develop.

The first observation is that our biological nature may *predispose* us to accept certain moral precepts, but it does not constrain us to accept them or to behave according to them. The same eminent intellectual abilities that make ethical behavior possible and necessary, and in particular free will, also give us the power to accept some moral norms and to reject others, independently of any natural inclinations. A natural predisposition may influence our behavior, but influence and predisposition are not the same as constraint or determination. It may be, then, that there are natural dispositions to selfishness (dispositions that Gilkey, Chapter 7, recalls the theologians have interpreted as original sin). But humans have the power to rise above these tendencies.

This observation deserves attention because authors such as Konrad Lorenz (1963) and Robert Ardrey (1966) have presented aggression and the territorial "imperative" as natural tendencies, which might therefore be futile to try to resist. Whether or not aggression and the territorial imperative are ingrained in our genes is neither obvious nor needs to be explored here. What needs to be said, however, is (1) that the morality of the behaviors in question is to be assessed in any case by the accepted norms of morality and not by recourse to biological evidence, and (2) that if such tendencies or imperatives did exist, people would still have the possibility and the duty of resisting them (even at the expense of a fitness reduction) whenever they are seen as immoral (Dobzhansky, 1973).

A second observation is that some norms of morality are consistent with behaviors prompted by natural selection, but other norms are not so. The commandment of charity, "Love thy neighbor as thyself," often runs contrary to the inclusive fitness of the genes, even though it promotes social cooperation and peace of mind. If the yardstick of morality were the multiplication of genes, the supreme moral imperative would be to beget the largest possible number of children and (with lesser dedication) to encourage our close relatives to do the same. But to impregnate the most women possible is not, in the view of most people, the highest moral duty of a man. Rather, the highest moral duties, as judged by cultural norms, have often been universal duties of justice and benevolence.

The third consideration is that moral norms differ from one culture to another and even "evolve" from one time to another. Today, many people see that the Biblical injunction "Be fruitful and multiply" has been replaced by a moral imperative to limit the number of one's children. No genetic change in human population accounts for this inversion of moral value. Moreover, an individual's inclusive fitness is still favored by having many children.

The evaluation of moral codes or human actions must take into account biological knowledge, but biology is insufficient for determining which moral codes are, or should be, accepted. This may be reiterated by returning to the analogy with syntactic language. Our biological nature determines the sounds that we can or cannot utter and also constrains human language in other ways. But a language's syntax and vocabulary are not determined by our biological nature (otherwise, there could not be a multitude of tongues), but are products of human culture. Likewise, moral norms are not determined by biological processes, but by cultural traditions and principles that are products of human history. That is the difference of being human.

References

Alexander, Richard D. 1979. *Darwinism and Human Affairs*. Seattle: University of Washington Press.

Ardrey, Robert. 1966. *The Territorial Imperative*. New York: Atheneum.

Ayala, Francisco J. 1980. *Origen y Evolución del Hombre*. Madrid: Alianza Editorial.

Ayala, Francisco J. 1982a. "The evolutionary concept of progress." In G. A. Almond et al., eds., *Progress and Its Discontents* (pp. 106–124). Berkeley: University of California Press.

Ayala, Francisco J. 1982b. "La naturaleza humana a la luz de la evolución." *Estudios Filosóficos* 31:397–441.

Ayala, Francisco J. 1987. "The biological roots of morality," *Biology and Philosophy* 2:235–252.

Barash, David P. 1977. *Sociobiology and Behavior*. New York: Elsevier.

Dobzhansky, Theodosius. 1973. "Ethics and values in biological and cultural evolution." *Zygon* 8:261–281.

Hamilton, William D. 1964. "The genetical evolution of social behavior." *Journal of Theoretical Biology* 7:1–51.

Hume, David. [1740]1978. *Treatise of Human Nature*. Oxford: Oxford University Press.

Huxley, Julian S. 1953. *Evolution in Action*. New York: Harper.

Huxley, Thomas H., and Julian S. Huxley. 1947. *Touchstone for Ethics*. New York: Harper.

Lorenz, Konrad. 1963. *On Aggression*. New York: Harcourt, Brace and World.

Moore, G. E. 1903. *Principia Ethica*. Cambridge: Cambridge University Press.

Ruse, Michael. 1984. "Review of Peter Singer, *The Expanding Circle*." *Environmental Ethics* 6:91–94.

Ruse, Michael. 1986a. *Taking Darwin Seriously: A Naturalistic Approach to Philosophy*. Oxford: Basil Blackwell.

Ruse, Michael. 1986b. "Evolutionary ethics: A phoenix arisen." *Zygon* 21:95–112.

Ruse, Michael, and Edward O. Wilson. 1986. "Moral Philosophy as Applied Science." *Philosophy: Journal of the Royal Institute of Philosophy* 61:173–192.

Spencer, Herbert. 1893. *The Principles of Ethics*. London.

Waddington, Conrad H. 1960. *The Ethical Animal*. London: Allen & Unwin.

Wilson, Edward O. 1975. *Sociobiology: the New Synthesis*. Cambridge, MA: Harvard University Press.

Wilson, Edward O. 1977. Foreword. In David P. Barash, *Sociobiology and Behavior*. New York: Elsevier.

Wilson, Edward O. 1978. *On Human Nature*. Cambridge, MA: Harvard University Press.

6

When Natural Selection and Culture Conflict

ELLIOTT SOBER

▪ ▪ ▪ *Editor's Introduction*

Ayala's byproduct is Elliott Sober's monkey wrench. Puzzling further over the question of the origin of the mind, Sober analyzes conflicts between the cultural sciences, where the conscious mind is a determinant of events, and biology, where it is not. "The human brain can throw a monkey wrench into the idea that adaptationism applies to human behavior with the same force that it applies to behaviors in other species. The brain is a problem for adaptationism because the brain gives rise to a process that can oppose the process of biological selection" (p. 156).

Sober finds that culture can and does override biology. "Biological selection produced the brain, but the brain has set into motion a powerful process that can counteract the pressures of biological selection. . . . Natural selection has given birth to a selection process that has floated free" (p. 158). That "floating free" is similar to what Margulis and Sagan intended by the "active forgetting" they urged when moral humans transcend their biological heritage. In this chapter, we get from a philosopher of biology, as we did from a geneticist in the last, an account that, if true, will require that Ruse's position be reformed. Ayala and Sober agree that human intelligence was targeted by natural selection because of its survival value; and both find, each with a different analogy, that the origin of mind introduces a critical turn into the evolutionary history. Natural selection, though it does not cease to operate, is transcended.

Distinguishing between cultural and genetic transmission processes, Sober is especially impressed with the tempo differential. "What becomes clear in these models is that in assessing the relative importance of biology and culture, *time is of the essence*. Culture is often a more powerful determiner of change than biological evolution because cultural changes occur *faster*" (p. 155).

If we consider the spread of ethical ideas, for instance, Sober notes that "it is not at all clear that evolutionary theory helps us explain why moral opinions about slavery changed so dramatically in nineteenth-century Western Europe" (p. 158). The explanation must rather lie in the sweeping changes in cultural ideas that came in so speedily. In a single century, three generations, we have little reason to think that the genetics of the European peoples changed at all. In the same century, there was a moral revolution; ideals of freedom and human rights prevailed.

Sober is reluctant to be metaphysical about this tempo difference. Presumably casting an eye toward Langdon Gilkey's insistence that, in a fuller account, "nature edges into mind; mind edges into nature" (Chapter 7, p. 168), and contradicting Charles Birch's "divine lure" (Chapter 8, p. 210), Sober warns, "The important thing is that cultural selection can be more powerful than biological selection. The reason for this is not some mysterious metaphysical principle of mind over matter. . . . The reason is humble and down to Earth: *thoughts spread faster than human beings reproduce*" (p. 156). Monkey wrench, yes, but mystery no. "Astonishing" (p. 151) yes, but metaphysical, no.

Sober's analysis brings considerable insight into why and how the human mind introduces novelties into the biological world. Something more emerged out of something less. Culture, B, follows biology, A, but develops its own differing logic, so that culture, B, is incommensurable (a monkey wrench) in the categories relevant to biology, A. There is a novel mode of rational transmission and selection. We may, however, begin to wonder whether there are yet deeper considerations than mere tempo of transmission when natural selection and culture conflict. If Sober is right, explanations are over. Nevertheless, the two subsequent authors will contend that there is something mysterious about matter that gives rise to mind, giving rise to a rational and ethical selection that is more powerful than biological selection.

We turn with interest to Sober's description of that possibility. Later, especially if we conclude that this possibility has become actual, we may wonder whether this humble, down-to-earth account of the astonishing arrival of higher values, including morality, is too much like possibilities floating in out of nowhere. "Monkey wrench" is an analogy in a way that "byproduct" is not. Are Sober's and Ayala's accounts of the emergence of morality consistent? Can cultural selection be more powerful than biological selection? Is the sole reason for this power the speed of transmission of ideas? The human mind is the "monkey wrench" that sets loose in culture factors not present in nonhuman natural selection, but what is the appeal of ideas that spread from mind to mind? In the example that Sober gives, why do Italian women come to desire fewer babies? What would eventually happen to such a population, reproducing differentially at two-fifths the rate of other populations? Would other populations of women also come to share this desire? In Sober's account, morality may float free from the pressures of natural selection, but where do the criteria for morality come from? What is the appeal that has persuaded persons that slavery is wrong? The idea spread rapidly, but why? Does more need to be said about the how ethical convictions originate and spread, superimposed on biological dispositions?

Elliott R. Sober is Professor of Philosophy at the University of Wisconsin, Madison. He is the author of *The Nature of Selection: Evolutionary Theory in Philosophical Focus* (1984), *Reconstructing the Past: Parsimony, Evolution, and Inference* (1988), *Philosophy of Biology* (1993), *Core Questions in Philosophy: A Text with Readings* (1991), and *Simplicity* (1975) and is the editor of *Conceptual Issues in Evolutionary Biology* (1984), as well as author of over eighty articles in the philosophy of science, especially the philosophy of biology. He has been a Guggenheim Fellow and received National Science Foundation and National Endowment for the Humanities Fellowships. •

1. *Border Disputes*

Scientific disciplines grow increasingly specialized and insulated from each other. This separation makes it tempting to think that disciplinary divisions carve nature at its joints. Since the physics department is separate from the biology department, and the biology department is separate from the history department, one might be inclined to think that physical, biological, and cultural phenomena are distinct from each other. But every so often, nature shows us that this supposition is naive. Nature has the habit of furnishing us with problems on which two apparently unrelated disciplines can be brought to bear. When this happens, the two disciplines often work together in reasonable harmony. For example, an evolutionist may want to date some fossils, so a physicist skilled in the techniques of carbon dating is consulted. However, sometimes the two disciplines come into conflict.

One historically important example of disciplinary conflict is Lord Kelvin's disagreement with Darwin about the age of the Earth. Darwin's hypothesis of slow and gradual evolution required an old Earth. Kelvin's physical theories put a more recent date on the Earth's origin. At the time, no biologist could say where Kelvin's calculations had gone wrong. But biologists convinced about Darwin's theory on other grounds felt sure that there must be a mistake somewhere. Only later, with the advent of theories of radioactive decay, did it become clear how this disciplinary conflict should be resolved. Kelvin, knowing nothing of radioactivity, had underestimated the Earth's age. Darwin's hypothesis of an old Earth was shown to be compatible with a revised geology.

A second historical example of disciplinary conflict can be found in the debate about continental drift. Long before the theory of plate tectonics was developed, biogeographers felt confident that South America and Africa must have been in contact. There is a detailed, point-for-point correspondence between the organisms found on one continent and the organisms found on the other. Migration could not explain this, nor could a land bridge. It is as if a piece of paper were torn down the center. The left half (South America) corresponds to the right half (Africa) in such a way that we are driven to the hypothesis that the two halves once were seamlessly joined. Geologists were skeptical of the hypothesis of continental drift until a great deal of geological evidence was amassed. Basically, they discounted or knew nothing of the biogeographical evidence. Only when the physical argument was worked out to their satisfaction did they accept the idea of continental drift.

Biologists often decry the allure of physics worship. When a biological argument comes into conflict with a physical one, why should we automatically think that the biology must be wrong? There is no rule inscribed in heaven that says that "harder" sciences are to be more trusted than "softer" ones. When biology and physics conflict, it is an open question as to which is correct.

It is easy for biologists to accept the observation that "harder" sciences do not always win when this idea serves to defend the integrity and importance of biology. However, my subject is not the relationship of physics and biology. Rather, I am interested in another border dispute, this one between some ideas in evolutionary biology and some ideas in the social sciences. When predictions about the effects of natural selection conflict with what historians or social scientists maintain, which of these should be trusted? Biologists need to take to heart the real lesson about not being seduced by the allure of physics. Just as the "harder" science can be wrong when physics conflicts with biology, so the "harder" science can be wrong when biology conflicts with the social sciences.

The conflict between evolutionists and social scientists that I want to consider concerns the controversy about sociobiology. Sociobiologists tend to look at behavior by applying the yardstick of survival and reproductive success. The working hypothesis about behaviors found in nonhuman species is that they are present because they conferred some sort of fitness advantage. When one observes an organism behaving in a certain way, the task is to figure out how that form of behavior is more advantageous than alternative behaviors, where advantage is calculated in the currency of surviving and reproducing. This research program, even when applied to nonhuman species, has been questioned because of its automatic invocation of natural selection. Since biological evolution can be caused by factors other than natural selection, why should we assume automatically that natural selection is the right framework to use in explaining an observed behavior? This is the dispute about *adaptationism*.

The same question, plus a new one, can be raised when the theory of natural selection is applied to a human behavior. Even if the behavior was produced by biological evolution, why think that natural selection is the only causal factor one needs to consider? However, an additional issue is said to arise when the organism doing the behaving is one of our own. Human beings are unique among living things because they have minds and culture. Mind and culture are supposed to throw a monkey wrench into the adaptationist's format for explaining behavior.

Evolutionists who are firmly convinced of the preeminence of natural selection when it comes to nonhuman species often give voice to this reservation. For example, Richard Dawkins (1976) describes the living world as evolving characteristics that help genes make copies of themselves.[1] However, when it comes to human beings, Dawkins, at the close of his book, finds that he has to make an exception. He says that the human mind has liberated us from the tyranny of our genes.[2] Some sociobiologists see this as a faint-hearted retreat. If human beings are the result of evolution, why make an exception of them? If adaptationism is a reasonable research strategy to use on the rest of the living world, why timidly abandon that research strategy when we turn the lens of inquiry onto ourselves?[3]

I want to explain why the human mind *is* a potential monkey wrench. I will not explore the question of whether adaptationism makes sense in nonhuman evolution.[4] Rather, I intend to clarify why a consistent adaptationism must be especially circumspect when it takes on the project of explaining human behavior.

Before giving my answer to the question of whether the human mind makes us an exception, I will try to provide a deeper appreciation of just why it is puzzling. Why should the human brain be such a big deal? Granted, it is a complicated organ. But so, in their own ways, are the stomach and the eye. It is entirely unclear why the complexity of the brain should exempt human beings from the adaptationist research program. After all, the human brain evolved. If natural selection is the principal cause of evolutionary change, then we should expect natural selection to be the principal cause of the brains we possess.

There is another way to appreciate the puzzle. Presumably, there are structural features of the brain that are prerequisites for the existence of language and culture. If so, we cannot explain the brain by appeal to language and culture. That would place the cart before the horse. Evolution produced the brain, which then gave rise to the various characteristics of human culture. We are here contemplating a causal chain with three links in it:

Evolution → The Human Brain → Culture

If causality is a transitive relation, we may conclude that an explanation of human culture is to be found in the facts of evolution. This seems enough to underwrite the sociobiological research program.

Looked at in this way, the social sciences are guilty of not pursuing explanatory problems with sufficient depth. They rest content with superficial cultural explanations of cultural phenomena. We need to push the problem farther back. To refuse to do so is to bury our heads in the sand or to deny the fact that we and our brains have evolved. If we neither deny nor ignore the fact of evolution, we seem bound to agree that human sociobiology is on the right track.

This argument, of course, does not show that any particular sociobiological explanation is correct. Maybe the specific explanations suggested to date are all erroneous.[5] The main argument we have to confront is at a much higher level of abstractness. It purports to show that our brains do not exempt us from the subject known as the evolution of behavior. The conclusion is that sociobiology is right *in general*, although it might be wrong so far about *the specifics*.

There is room to quibble with this argument with respect to the issue of transitivity. Karl Marx wrote *Capital*. I use that three-volume book as a paper weight. As a result, the wind does not blow my papers around. We have here a causal chain with three links, from Marx to *Capital* to my sedate papers. However, it seems odd to say that Marx causes my papers to remain in place.

Although general questions about the logic of explanation and causality are interesting, I will not pursue them here. Maybe explanation and causation are not transitive *in general*. Be that as it may, there is nonetheless a special feature of how natural selection works that prevents us from using that fact to dismiss the defense of sociobiology I have sketched.[6]

2. *Proximate Mechanisms and Ultimate Causes*

Ernst Mayr (1961) introduced a useful distinction between two types of explanation that biologists construct. Physiologists, morphologists, and students of behavior construct *proximate explanations*. Evolutionists, on the other hand, construct what Mayr called *ultimate explanations*. The difference can be grasped by considering the following question: Why do ivy plants grow toward the light? A plant physiologist would answer this question by describing physical structures internal to ivy plants that cause them to grow toward the light. An evolutionist, on the other hand, would answer the question by describing why the trait of phototropism evolved. If the evolutionist's answer is formulated in terms of natural selection, it will tell us why phototropism conferred a selective advantage on plants in the lineage leading up to the ivy plants we observe.

There is no conflict between these two modes of explanation. Proximate explanations describe ontogenetic causes; ultimate explanations describe phylogenetic causes. The plant physiologist cites causes that exist within the lifetime of the individual ivy plant. The evolutionist formulates explanations that have a longer time frame. Once again, we have a causal chain with three links:

$$\text{Natural Selection} \rightarrow \text{Physical Structure } X \rightarrow \text{Phototropism}$$

Natural selection has caused presently existing plants to possess physical structure X, and that physical structure causes the behavior we call phototropism. The impulse to invoke the transitivity of causality is hard to resist. Natural selection causes phototropism.

There is a special wrinkle in causal chains of this sort that is worth noting. The reason natural selection led the plant to possess physical structure X is that X causes phototropism. The internal structure is selected precisely because of the behavior it produces. This is not true for the causal chain involving Marx, *Capital*, and my papers. Marx did not write *Capital* because of the effect that *Capital* has (or would have) on papers not being blown by the wind.

Whenever natural selection causes a behavior to evolve, it must equip the organism with internal machinery that will produce the behavior. In the present example, for natural selection to cause phototropism to evolve, it must furnish plants with an internal mechanism that causes them to grow toward the light. This mechanism may be thought of as having two functional parts. It must contain a *detector* and an *effector* (Williams, 1966). The detector gives the plant a way to tell from what direction the light is coming. The effector causes the plant to grow in that direction. In principle, there might be different internal mechanisms that could achieve the desired behavior. Natural selection will cause one or more of these to evolve. These internal mechanisms are the proximate devices that exist because they get the organism to behave in the right way in the right circumstances.

It goes without saying that plants are able to grow toward the light without having *minds*. They do not form beliefs about where the light source is, nor do they desire to grow in that direction. Detectors and effectors do not have to be mental mechanisms. This is one important difference between humans and plants. Like plants, we possess internal mechanisms that cause us to produce different behaviors in different circumstances. But unlike plants, our detectors and effectors involve beliefs and desires. Unlike physical structure X, the human brain provides us with *minds*, which cause us to behave in the ways we do.

If we think of the brain in this fashion, it becomes difficult to see why the mind (or the brain) should be able to "liberate us from the tyranny of the genes." Our brains are proximate mechanisms, which are related to our behavior in the same way that the internal mechanisms in a plant are related to its behavior. If ivy plants could talk, their self-love might lead them to proclaim that they are not slaves to the genes they contain. Perhaps they would seize on some characteristic unique to ivy plants and say that it exempts ivy plants from usual patterns of evolutionary explanation. But, of course, ivy plants can't talk; we can. Granted, the human mind/brain is unique. Why does this feature show that our behaviors are not explicable by evolution and natural selection?

3. *Lability*

One suggestion is that the human mind exempts us from usual patterns of evolutionary explanation because human behavior is so *labile*. We have enormous behavioral flexibility; ivy plants do not. Because of this, the behaviors of ivy plants, and the internal mechanisms that produce them, fall within the scope of selectionist explanations, while the behaviors of human beings, and the mental characteristics that cause them, do not.

To evaluate this line of argument, we must ask what it means to say that a behavior is "labile." The distinction we need is between what biologists call *obligate* and *facultative* traits. An obligate trait is unconditional; the organism has the trait, no matter what its environment is like. Facultative traits are conditional; the organism displays them in some circumstances, but not in others. To some degree, whether we classify a behavior as obligate or facultative depends on how we describe it. It is true that ivy plants *always grow toward the light*. Phototropism is hard-wired. Described in this way, the behavior is obligate. But we also can describe the ivy plant as growing in a given direction *if there is light there*. Phototropism means that the direction of growth is conditional on a light gradient. Described in this way, the behavior is facultative.

This same relativity applies to behaviors that are uniquely human. I *always avoid reading science fiction*. This is an unconditional fact about my behavior. However, it also is true that I avoid reading a book *if it is science fiction*. Here I describe my behavior so that it is facultative, not obligate. To say that a behavior is labile is simply to say that it is facultative; its occurrence is conditional on the presence of various environmental factors. A moment's thought shows that evolutionary explanations of behaviors in other species often focus on labile behaviors. Plants grow in some directions, but not others; bees visit some flowers but not others; a peacock courts some organisms but not others. Conditional behaviors are grist for the mill of selectionist explanation.

It may be suggested that ivy plants are programmed to grow toward the light, but that we are not programmed to behave in the way we do. If we could make sense of this idea, maybe we would find a key to understanding why human behavior falls outside the purview of evolutionary explanation.

I have my doubts about this suggestion. What does *programmed* mean? One natural interpretation is that *programmed* means *caused*. When we say that ivy plants are programmed to seek the light, we mean that they possess internal mechanisms that cause them to do so. However, if we accept the assumption that human behavior is caused by the subject's mental states, then we do not have here a difference between ivy and us, but a similarity.

A second interpretation is possible: ivy plants do not have free will, but we do. Maybe this is what it means to say that they, but not we, are "programmed" to behave in certain ways. I am inclined to agree that we have free will but plants do not. However, I do not think that this solves the problem at hand. The real issue is whether our behaviors are caused. If they are, then they seem as subject to evolutionary explanation as the behavior of ivy plants.

Philosophers standardly distinguish two approaches to the relation of causality and free will. *Compatibilists* such as myself say that it is possible for a behavior to be free and caused at the same time. *Incompatibilists* say that causality rules out freedom. Although I do not have time to argue that compatibilism is correct, it is worth bearing in mind that science has lots of evidence that supports the claim that behavior, ours included, is caused; there is no evidence whatever that supports the opposite claim.[7] For this reason, I think a sensible starting point for discussion of free will is the assumption that our behavior is caused. Once this is granted, we can ask whether freedom also may be accommodated within a scientific world picture.

Although the issue of freedom, viewed in this way, is an interesting one, it is not germane to the problem I am considering. Our brains cause us to produce various behaviors. It does not matter whether those behaviors have the additional property of being produced by a free will. Of course, the behaviors we produce are labile; the environment influences which behaviors we produce in which circumstances. This is enough to make the relation of the mind to the behaviors it produces look very much like the relation between the proximate mechanisms and the behaviors we find in other species. Lability, per se, gives us no basis for exempting a behavior from evolutionary explanation.

4. *Nature Versus Nurture*

Sometimes the argument for thinking that human sociobiology must be wrong, even if nonhuman sociobiology is correct, is based on the distinction between nature and nurture. The rest of the living world does what it does by nature; we act as we do because of nurture. Translated into modern language, this means that other species are caused to act as they do by their genes, whereas we are caused to act as we do by our cultural environment.

To understand how hypotheses about natural selection are related to the nature/nurture controversy, we must take care to distinguish two stages in the evolution of a trait. We must distinguish the trait *while it is on its way to becoming universal* in the population and the trait *once it has reached 100 percent representation* (fixation). Natural selection produces evolution only if the traits under selection are differentially *heritable*. But this says nothing about what is true after the favored trait has reached fixation. Fixed traits are always inherited, and there is nothing for natural selection to select, since all individuals have the same trait.

Consider, for example, how an evolutionist would explain the opposable thumb. It might be suggested that this trait evolved because opposable thumbs provided some advantage in survival; maybe opposable thumbs enhanced the organism's ability to create and manipulate tools. If selection is to account for the trait's prevalence, we must postulate an ancestral population in which some individuals had the trait while others did not. Individuals with opposable thumbs fared better in the struggle for existence, and so the trait increased in frequency. For this to happen, people with the trait must have been genetically different from people without it. Without this condition, the trait would not have been differentially heritable (in the technical, "narrow" sense of *heritability*), and differential heritability is a prerequisite for selection to change the composition of the population.

So a selectionist explanation of the opposable thumb requires us to think that there *was* a genetic difference between individuals with opposable thumbs and individuals without such thumbs *while the trait was evolving*. However, matters change when we look at the population after the trait has finished evolving. When we survey current human populations, we find that not everybody has an opposable thumb. But this is not true any longer because thumbs are differentially heritable. Some people have experienced industrial accidents; others have mothers who took the drug Thalidomide. The point to recognize is that *current* variation in the trait may be mainly or exclusively environmental. It is quite possible that there is no genetic difference between individuals *now* who have the trait and individuals *now* who do not. Evolution by natural selection requires heritability while the trait was evolving; it does not require that the trait remains differentially heritable after it has become fixed.

I conclude that even if it were true that the variation in behavior that we *now* observe among human beings is mainly or exclusively environmental, this fact would not exempt human behavior from evolutionary explanation.

There is a second reason why an environmentalist position on the nature/nurture dispute does not exempt human behavior from explanation in terms of natural selection. We need to realize that there are different approaches within the sociobiological research program, and that one prominent approach actually embraces an extreme environmentalism.

The writings of Edward O. Wilson (1975, 1978) have made it easy to associate sociobiology with the assumption of "biological determinism." Although Wilson rejects this characterization of his own work, the fact remains that he often postulates genetic differences to account for behavioral differences. Consider, for example, his discussion of behavioral differences between the sexes. Wilson thinks that natural selection has helped shape such behavioral differences and that genetic differences help account for why boys tend to act one way and girls another. The same pattern can be found in his discussion of homosexuality; he talks about a gene for homosexuality and how it might have evolved.

A quite different approach to behavioral differences among human beings can be found in the work of Richard Alexander (1979, 1987). Alexander emphasizes the enormous behavioral lability that our species possesses. He regards this ability to adjust behavior in the light of environmental conditions as universal within our species and unique to it. When Alexander finds that some people behave differently from others, he does not conjecture that this is due to genetic differences. Rather, he conjectures that the individuals live in different environments. His guiding hypothesis is that human beings adjust their behaviors so that the behavior they produce in a given environment is the one that would maximize fitness in that environment. People in different societies act differently, he would say, because what enhances survival and reproduction in one environment may not do so in another. In short, Alexander is a radical environmentalist when it comes to explaining behavioral differences among human beings. His position is about as far from "biological determinism" (which in common parlance means *genetic* determinism) as one can get.

In short, human sociobiology would not be stopped dead by the fact, if it is a fact, that behavioral variation within our species is mainly environmental rather than genetic. I have given two reasons for saying this. Now I will add a third. We must be careful to distinguish the problem of explaining variation within a species from the problem of explaining variation between species. Even if within-species variation is largely environmental, it still may be true that between-species variation is largely genetic. Human sociobiology may choose to focus on the first type of variation, on the second, or on both. So even if within-species variation could not be handled in the sociobiological research program, nothing follows about sociobiology's ability to explain why we humans differ from the rest of the living world. Presumably, genes do play an important role in explaining why human behavior differs from the behavior of chickens. It is entirely unclear why the fact that we have a brain exempts us from an evolutionary explanation of this difference.

The question I am posing is why an adaptationist about the rest of nature might have reason to suspect that adaptationism is problematic in the case of human beings. The idea that the nature/nurture distinction helps underwrite this position deserves to have one last nail driven into its coffin. Maybe it is true that lots of behavioral variation among human beings is environmental rather than genetic. The point I would emphasize is that the same is probably true for lots of traits in lots of other species. If a foraging animal has been programmed by natural selection to choose the optimal food item available in its environment, then it will be no surprise that animals in the valley have a different diet from those of the same species that live in the hills. They behave differently because their environments differ. Far from embarrassing adaptationism, this is just the sort of fit between organism and environment that adaptationists relish.

5. *How to Oppose Natural Selection*

Having catalogued a number of false starts, I now want to say why the human mind/brain does pose a serious impediment to adaptationism. I want to show why sociobiology applied to human beings has a smaller chance of being successful than sociobiology applied to the rest of nature. The point is not to find some difficulty with selectionist explanations *in general*, but to show why they are problematic in the *specific* case of human behavior. We must see that the brain triggers a process that is quite different from the processes initiated by the eye or the stomach. By this I do not mean that we must examine how the nerve impulses sent by the brain to the rest of the body differ from the gastric juices that the stomach unleashes on our food. Rather, we must understand how the brain causes a *social* process. Brains influence brains in a way quite unlike the way stomachs influence stomachs, or eyes influence eyes.

The astonishing thing about the human brain is that it has brought into being a selection process of its own. The brain is able to liberate us from the control of biological evolution precisely because it has given rise to the opposing process of cultural evolution.

To flesh out this idea, we first must understand what the process of cultural selection actually means. After that, the task is to explain how this different kind of selection process is able to counteract the effects of biological evolution. Even if cultural evolution opposes the tendencies of biological evolution, why should we expect cultural evolution to triumph? Instead, why not think that when biology conflicts with culture, it is the biological process that always wins?

We need to see how biological selection and cultural selection are similar, and also how they differ. To this end, it will be useful to describe three formats to which a selection process may conform. As will become clear, I will be using *selection* as well as other terms with meanings that are broader than is customary in evolutionary theorizing.

Given a set of objects that exhibit variation, what will it take for that ensemble to evolve by natural selection? By evolution, I mean that the frequency of some characteristic in the population changes. Two ingredients are crucial. The first is that the objects differ with respect to some characteristic that makes a difference in their abilities to survive and reproduce. Second, there must be some way to insure that offspring resemble their parents. The first of these ingredients is differential *fitness*; the second is *heritability*.

In most standard models of natural selection, offspring resemble their parents because a genetic mode of transmission is in place. And traits differ in fitness because some organisms have more babies than others, babies that survive to reproductive age. It may seem odd to say that "genes" are one way to produce heritability and "having babies" is one way to measure fitness, as if there could be others. My reason for saying this will soon become clear.

The second form that a selection process can take retains the idea that fitness is measured by how many babies an organism produces, but drops the idea that the relevant phenotypes are genetically transmitted. Strictly speaking, evolution by natural selection does not require genes. It simply requires that offspring resemble their parents. For example, if characteristics were transmitted because children imitated their parents, a selection process could occur without the mediation of genes.

A hypothetical example of how this might happen can be constructed by considering the incest taboo. Suppose that incest avoidance is advantageous because individuals with the trait have more viable offspring than individuals without it. If offspring *learn* to be incest avoiders from their parents, the frequency of the trait in the population may evolve. This can occur without there being any genetic differences between those who avoid incest and those who do not (Colwell and King, 1983).

In this second kind of selection model, mind and culture displace one but not the other of the ingredients found in models of the first type. In the first sort of model, a genetic mode of transmission works side by side with a concept of fitness defined in terms of reproductive output—what I have called "having babies." In the second, reproductive output is retained as the measure of fitness, but the genetic mode of transmission is replaced by a psychological one. Teaching can provide the requisite heritability just as much as genes.

The third pattern for applying the idea of natural selection abandons both of the ingredients present in the first. Genes are discarded as the mode of transmission. And fitness is not measured by how many babies an organism has. Individuals acquire their ideas because they are exposed to the ideas of their parents, of their peers, and of their parents' generation; the transmission patterns may be vertical, horizontal, and oblique.

An individual exposed to this mix of ideas need not give them all equal credence. Some may be more attractive than others. If so, the frequency of ideas in the population may evolve. Notice that there is no need for organisms to differ in terms of their survivorship or degree of reproductive success in this case. Some ideas catch on while others become passé.

It is evident that this way of modelling cultural change is no more tied to the genetic theory of natural selection than it is tied to epidemiology. Rumors and diseases exhibit a similar dynamic. The spread of a novel characteristic in a population by natural selection, like the spread of an infection or an idea, is a diffusion process.[8]

Examples of this third type of selection process are discussed in different versions of evolutionary epistemology. Karl Popper (1973) suggested that scientific theories compete with each other in a struggle for existence. Better theories spread through the population of researchers; inferior ones exit from the scene. Popper highlighted the nonbiological definition of fitness used in this view of the scientific process when he said that "our theories die in our stead." Other models in evolutionary epistemology are structured similarly (Campbell, 1974; Hull, 1988).[9]

The three forms a selection process can take are summarized in the following table:

Three Types of Selection Process

	Heritability	*Fitness*
I	genes	having babies
II	learning	having babies
III	learning	having students

In this taxonomy, "learning" should be taken broadly; it does not require anything very cognitive, but can simply involve imitation. The same goes for "having students"; all that is involved is successful influence mediated by learning.

The parallelism between Types I and III is instructive. In Type I processes, individuals produce different numbers of babies in virtue of the phenotypes they have (which are transmitted genetically); in Type III, individuals produce different numbers of students in virtue of the phenotypes they have (which are transmitted by learning).[10]

I do not insist that this classification is exhaustive. Arguably, the spread of a contagious disease falls happily into none of these three categories (that is, unless the concept of "learning" is stretched considerably). Still, I hope that it is clear how the three categories differ.

Type I selection processes give rise to biological evolution; Type III processes cause cultural evolution. Type II is a kind of intermediate hybrid. In principle, each type of process can influence the evolution of mind and culture. For example, it is conceivable that ideas about incest avoidance have evolved because of each of these three processes. The difference between biological and cultural selection does not concern *what* traits they affect, but *how* those traits are made to evolve.

With biological and cultural selection distinguished in this way, we now can ask what happens to a characteristic that is simultaneously influenced by both. Cavalli-Sforza and Feldman (1981) and Boyd and Richerson (1985) have developed a number of models in which Type I and Type III processes are both considered. I now describe one of the Cavalli-Sforza/Feldman examples.

In the nineteenth century, Western societies exhibited an interesting demographic change, one that had three stages. First, oscillations in death rates due to epidemics and famines became both less frequent and less extreme. Second, overall mortality rates steadily declined. The third part of this demographic transition was a dramatic decline in birth rates.

Consider the last part of this transformation. Why did fertility decline? From the point of view of a narrowly Darwinian outlook, this change is puzzling. A characteristic that increases the number of viable and fertile offspring will spread under natural selection, at least when that process is conceptualized from the point of view of a Type I model. Cavalli-Sforza and Feldman are not tempted to appeal to the theory of optimal clutch size, following Lack (1954), according to which a bird parent sometimes can augment the number of offspring surviving to adulthood by laying fewer eggs. They can better rear fewer nestlings. Presumably, this Darwinian suggestion is not worth exploring in the present case, because women in nineteenth-century Western Europe could have had more viable offspring than they did. People were not caught in the bind that Lack attributed to his birds.

The trait that increased in the modern demographic transition entailed a reduction in biological fitness. The trait spread *in spite of* its being biologically disadvantageous. In Italy, women changed from having about five children on average to having about two. The trait of having two children, therefore, has a biological fitness of 2/5, when compared with the older trait it displaced.

Cavalli-Sforza and Feldman emphasize that this demographic change could not have taken place if traits were passed down solely from mothers to daughters. This point holds true whether characteristics are genetically transmitted or learned from mother to daughter. A woman with the new trait will pass it along to fewer offspring than a woman with the old pattern, if a daughter is influenced just by her mother.

What is required for the process is some mixture of horizontal and oblique transmission. That is, a woman's reproductive behavior must be influenced by her peers and by her mother's contemporaries. However, it will not do for a woman to adopt the behavior she finds represented on average in the group that influences her. What is required is that women find small family size more attractive than large family size even when very few peers possess the novel characteristic. There must be a "transmission bias" in favor of the new trait.

Having a small family was more attractive than having a large one, even though the former trait had a lower Darwinian fitness than the latter. Cavalli-Sforza and Feldman show how the greater attractiveness of small family size can be modelled by adapting mathematical ideas from evolutionary biology. However, when these biological ideas are transposed into a cultural setting, one is talking about cultural fitness, not biological fitness. The model they construct of the demographic transition combines two selection processes. When fitness is defined in terms of having babies, there is selection *against* having a small family. When fitness is defined in terms of the attractiveness of an idea, there is selection *favoring* a reduction in family size. Cavalli-Sforza and Feldman show how the cultural process can overwhelm the biological one. If the trait is sufficiently attractive (and their models have the virtue of giving this idea quantitative meaning), the trait can evolve in spite of its Darwinian disutility.

One virtue of this and related models of cultural evolution is that they place culture and biology into a common framework so that their relative contributions to an outcome are rendered commensurable. What becomes clear in these models is that, in assessing the relative importance of biology and culture, *time is of the essence*. Culture is often a more powerful determiner of change than biological evolution because cultural changes occur *faster*. When biological fitness is calibrated in terms of having babies, its basic temporal unit is the span of a generation. But think how many replication events can occur in that temporal interval when the reproducing entities are ideas that jump from head to head. Ideas spread so fast that they can swamp the slower (and hence weaker) impact of biological natural selection.

Factors that affect evolution are stronger or weaker, according to the amount of change they bring about *per unit of time*. A process that increases a gene from 5 percent to 20 percent representation in a population within a single generation is more powerful than a process that takes fifty generations to achieve the same result. Biological selection, as Darwin suggested, often works on the basis of modest differences in fitness and so takes many generations to change significantly the composition of a population. Cultural evolution, evidently, often works much faster. This is what makes it powerful.

There is a vague idea about the relation of biology and culture that these models of cultural evolution help lay to rest. This is the idea that biology is "deeper" than the social sciences, not just in the sense that it has developed farther, but in the sense that it investigates more powerful causes. This idea involves a prejudice in favor of the causes described by the "harder" science. It asserts that if Darwinian selection favors one trait, but cultural influences favor another, the deeper influence of biology must overwhelm the more superficial influence of culture. Cavalli-Sforza and Feldman and Boyd and Richerson deserve credit for showing why this common opinion rests on a confusion.[11]

6. *The Application to Ethics*

We now have an answer to the question of why the human brain can throw a monkey wrench into the idea that adaptationism applies to human behavior with the same force that it applies to behaviors in other species. The brain is a problem for adaptationism because the brain gives rise to a process that can oppose biological selection. And it is not just that cultural selection can find itself at odds with biological selection. The important thing is that cultural selection can be more powerful than biological selection. The reason for this is not some mysterious metaphysical principle of mind over matter. When cultural selection is more powerful than biological selection, the reason is humble and down to earth: *thoughts spread faster than human beings reproduce.*

It is a standard idea in evolutionary theory that an organ will have characteristics that are not part of the causal explanation of why the organ evolved. The heart makes noise, but that is not why the heart evolved. Rather, the heart evolved because it pumps blood. We might mark this difference by saying that the function of the heart is to pump blood; its function is not to make noise. Making noise is a *side effect*; it is an evolutionary *spinoff*.

We must not lose sight of this distinction when we consider the human mind/brain. The organ evolved because of some of the traits it has. However, this should not lead us to expect that *every* behavior produced by the human mind/brain is adaptive. The mind/brain presumably has many *side effects*. It generates thoughts and feelings that have nothing to do with why the brain evolved.

Both brains and hearts have adaptive features and features that are evolutionary side effects. But to this similarity a fundamental difference must be added. When my heart acquires some characteristic (e.g., an improved circulation because I have taken up jogging), there is no mechanism in place that can cause that feature to spread to other hearts. In contrast, a thought—even one that is neutral or deleterious with respect to my survival and reproduction—is something that may expand beyond the confines of the single brain that produced it. Ideas can plug into a network in which brains are linked to each other by relations of mutual influence. This is a confederation that our brains have effected, but our hearts have not.

How might this account of the relation of biological and cultural evolution apply to ethics? Evolution has produced the human mind/brain, and that organ has made it possible for us to have a morality. However, this leaves open *why* the capacity for morality evolved and *why* we have the particular moral convictions we do. The sociobiological viewpoint advocated by Ruse (1986 and Chapter 4) asserts that the capacity for morality and the specifics of what we take to be moral are adaptations. They exist because they promoted survival and reproductive success. In contrast, Ayala (Chapter 5) argues that our capacity for ethics is evolutionary spinoff (on which see Sober, 1988, as well), while the content of our morality is the result of cultural, not biological, evolution.

My argument here is less ambitious than Ruse's and Ayala's. I have not proposed an explanation of either the content of our moral views or of our capacity for having a morality. Rather, I have tried to describe a *possibility*. I have tried to show why the brain is able to do something that other adaptive organs cannot do. Whether it has *actually* done so is another matter, one that I have not addressed in any detail.

The one example I discussed was the demographic transition. However, it would be a mistake to think that what is true in this case must be true in all others. When biological selection and cultural selection oppose each other, it is an open question which will be more powerful. No simple recipe can be provided; everything depends on the details.

When it comes to applying this set of ideas to the problem of explaining morality, it is important to break the phenomenon we call "morality" into pieces. Rather than asking whether "our morality" is the product of natural selection, we should focus on some specific aspect of that morality. Perhaps there is a simple evolutionary explanation for why no society demands universal infanticide. On the other hand, it is not at all clear that evolutionary theory helps us explain why moral opinions about slavery changed so dramatically in nineteenth-century Western Europe. Rather than look for some sweeping global "explanation of morality," it might be better to proceed piecemeal.

The idea that cultural evolution can swamp biological evolution does not imply that standard processes of biological evolution no longer operate at all in our species. Individuals still live and die differentially, and differential mortality often has a genetic component. This biological process is not *erased* by the advent of mind and culture. The biological process remains in place, but is joined by a second selection process that is made possible by the human mind.

It is quite true that biological evolution produced the brain, and that the brain is what causes us to behave as we do. However, it does not follow from this that the brain plays the role of a passive proximate mechanism, simply implementing whatever behaviors happen to confer a Darwinian advantage. Biological selection produced the brain, but the brain has set into motion a powerful process that can counteract the pressures of biological selection. The mind is more than a proximate mechanism for the behaviors that biological selection has favored. It is the basis of a selection process of its own, defined in terms of its own measures of fitness and heritability. Natural selection has given birth to a selection process that has floated free.

Acknowledgment

My thanks to the participants at the Conference on Biology, Ethics, and the Origins of Life at Colorado State University for useful discussion.

References

Alexander, Richard D. 1979. *Darwinism and Human Affairs*. Seattle: University of Washington Press.

Alexander, Richard D. 1987. *The Biology of Moral Systems*. New York: Aldine de Gruyter.

Boyd, Robert, and Peter J. Richerson. 1985. *Culture and the Evolutionary Process*. Chicago: University of Chicago Press.

Campbell, Donald T. 1974. "Evolutionary epistemology." In P. A. Schilpp, ed., *The Philosophy of Karl Popper*. LaSalle, IL: Open Court Publishing Co.

Cavalli-Sforza, L. L., and M. Feldman. 1981. *Cultural Transmission and Evolution: A Quantitative Approach*. Princeton, NJ: Princeton University Press.

Colwell, R., and M. King. 1983. "Disentangling genetic and cultural influences on human behavior—problems and prospects." In D. Rajecki, ed., *Comparing Behavior: Studying Man Studying Animals*. Hillsdale, NJ: L. Erlbaum Publishers.

Dawkins, Richard. 1976. *The Selfish Gene*. New York: Oxford University Press.

Gould, Stephen J., and Richard C. Lewontin. 1979. "The spandrels of San Marco and the Panglossian paradigm: A critique of the adaptationist programme." *Proceedings of the Royal Society of London, B. Biological Sciences* 205:581–98.

Hull, David L. 1988. *Science as a Process*. Chicago: University of Chicago Press.

Kitcher, Philip. 1988. *Vaulting Ambition: Sociobiology and the Quest for Human Nature*. Cambridge, MA: The MIT Press.

Lack, David. 1954. *The Optimal Regulation of Animal Numbers*. Oxford: Oxford University Press.

Maynard Smith, John. 1978. "Optimization theory in evolution." *Annual Review of Ecology and Systematics* 9:31–56.

Mayr, Ernst. 1961. "Cause and effect in biology." *Science* 134:1501–1506.

Orzack, Steven H., and Elliott Sober. (1994). "Optimality models and the test of adaptationism." *The American Naturalist* 143:361–380.

Popper, Karl R. 1973. *Objective Knowledge*. Oxford: Oxford University Press.

Ruse, Michael. 1986. *Taking Darwin Seriously*. Oxford: Basil Blackwell.

Sober, Elliott. 1984. *The Nature of Selection*. Cambridge, MA: MIT Press.

Sober, Elliott. 1988. "What is evolutionary altruism?" In B. Linsky and M. Matthen, eds., *New Essays on Philosophy and Biology* (pp. 75–99). University of Calgary Press.

Sober, Elliott. (1993). *Philosophy of Biology*. Boulder, CO: Westview Press.

Williams, George. 1966. *Adaptation and Natural Selection*. Princeton: Princeton University Press.

Wilson, Edward O. 1975. *Sociobiology: The New Synthesis*. Cambridge, MA: Harvard University Press.

Wilson, Edward O. 1978. *On Human Nature*. Cambridge, MA: Harvard University Press.

Endnotes

1. I develop objections to Dawkins' theory of the selfish gene in Sober (1984) and in Sober (1993). However, these objections do not concern Dawkins' adaptationism or his idea that the human mind radically alters the bearing of natural selection on the task of explaining human behavior.

2. Gilkey (Chapter 7) argues that there is a contradiction between Dawkins' view of biological evolution and Dawkins' liberal and altruistic ethical convictions. I do not think that there is a contradiction here. If Dawkins' biology entailed that it is *impossible* for us to be psychologically altruistic, then his biology would contradict the claims of an altruistic ethics (via the ought-implies-can principle). However, Dawkins does not believe that his biology has that implication. I hope in what follows to explain why Dawkins is correct in saying this. For discussion of the difference between evolutionary and psychological altruism and egoism, see Sober (1988).

3. In his essay in the present volume, and his earlier book (Ruse 1986), Ruse argues that our ethical beliefs are *adaptive illusions*. At least in the case of ethics, Ruse does not believe that mind and culture fundamentally alter the format one should use in explaining behavior. I discuss Ruse's idea that evolution shows that our ethical beliefs are false in Chapter 7 of Sober (1993).

4. The initial issues in this dispute were posed by Gould and Lewontin (1979) and Maynard Smith (1978). For further discussion, see Orzack and Sober (1994).

5. Kitcher (1984) develops a number of criticisms of specific sociobiological explanations (most of them pertaining to human beings), but does not contend that the subject cannot improve. Quite the contrary; since his objections are specific rather than at the level of first principles, he grants that evolutionary accounts of behavior, not exempting human beings, are perfectly possible.

6. There is another reason to be skeptical about this argument that attempts to show that evolution has interesting things to say about human behavior. The same style of argument would allow one to conclude that *particle physics* can explain human mind and culture; after all, the brain has the psychological capacities it does in virtue of its physical organization and this physical organization can be explained, ultimately and in principle, by physics. I leave it to the reader to consider what conclusion should be drawn from this parallelism. If sociobiology makes sense, does the same hold true of sociophysics?

7. I do not think that the failure (to date) to find a causal explanation for some behavior is evidence that the behavior has no cause. Also, I am not denying that the causes of human behavior differ in fundamental ways from the causes of phototropism.

8. This kind of selection process is what Dawkins (1976) had in mind in his discussion of "memes."

9. Even if the success of scientific ideas is influenced by nonevidential factors (e.g., political ideology or metaphysical palatability), ideational change in the

population of inquirers can be modelled as a selection process. See Sober (1993) for further discussion.

10. Notice that models of Type III selection processes do not require a genotype/phenotype distinction, nor do they demand that variation be generated randomly. These possible disanalogies between biological and cultural evolution do nothing to undercut the promise of evolutionary epistemology.

11. In Sober (1993), I argue that these models of cultural evolution will have a rather limited relevance to explanatory problems in the human sciences. The reason is that these models focus mainly on the *consequences* of transmission biases, whereas a historian will be interested in the *sources* of those biases. However, this point does not undermine the importance of these models with respect to the problem of showing how biological and cultural evolution can interact.

7

Biology and Theology on Human Nature

Ethics and Genetics

LANGDON GILKEY

■ ■ ■ *Editor's Introduction*

Langdon Gilkey, a theologian celebrated for his work in science and religion, pushes the revelatory character of the human phenomenon even further. The beginning and middle of the story, the biological chapters, have to be interpreted in the light of the end to which they lead. If culture must be interpreted as the outcome of biology, biology must equally be interpreted as the precursor of culture. When biology follows geochemistry, and life originates from nonbiotic materials, whence the possibilities become actual? When social science follows biology, do we expect culture to be the outcome of nature, explained in natural categories, or do we see biology as necessary but not sufficient to explain the social sciences?

Gilkey nowhere denies the human origins in biology, either in the past or continuing in the present. "*All* our possibilities are in the end genetic; our physical, psychological, moral, and spiritual similarities, as well as all our differences, have their patterns set in our genetic endowments. Thus it is true that our common social customs, manners, laws, morals, mores, and our religious rites, myths, and laws, can, or could be, traced back into the dim recesses of the human past. Our behaviors and beliefs evolve no less than our bones and our biochemistries" (p. 168). That, he reminds us, is as true of science as of ethics or religion.

But origins do not always yield the whole of meanings. "We are in part self-directing, centered beings. . . . Strangely, we humans must choose, affirm, and use these possibilities of our nature. . . . The call to 'choose ourselves,' to embody in our existence what is given to us, to embody it by inner decision, affirmation, commitment, and perseverance, is one of the characteristics that makes us human; it is the source of responsibility, and so of the moral. . . . Morals and science alike spring from the *humanum*, what theology has classically termed 'the image of God'" (p. 169).

Stretching the present back into the biological past affects our vision of nature as much as it affects our vision of culture. "Here nature as 'matter'—nature as known by the physical sciences, as *reduced*, one might say—shows its deeper, and more mysterious, identity with nature as *also* the source of psyche and of spirit" (p. 170). Gilkey's conclusion contrasts, rather starkly, with Sober's refusal to find mysterious the appearance of mind, or ethics. Nature, revealed as the source of spirit, enlarges the metaphysical picture. "Nature edges into mind; mind edges into nature" (p. 168).

Gilkey reinforces this general conclusion by turning to press the sociobiological accounts of human nature against the sociobiologists themselves. Their accounts fall into an incoherent dualism. They discover, what theology already knew, that the human condition is broken. But they have no provision for redeeming this broken condition, for moving to the nobler chapters of the human story. Nevertheless, the experience of morality, choosing the right against the wrong, no less that choosing the wrong against the right, is evidently among the phenomena of human experience. Morality does somehow manage to arise out of this broken human condition, and for this biologists have no explanation. They have to speak of a "byproduct" (Ayala) or of a "monkey wrench" (Sober). Or they urge "rebellion" against our "selfish" genes (Dawkins), twisting analogies as though genes could be experientially selfish and a mere call to "Rebel!" could redirect human nature.

Biologists regularly not only urge, but themselves illustrate, the achievement of a nonselfish morality. They inhabit the transcending ethical, evaluative, responsible human world experientially while professing to be reductionists biologically. Gilkey finds paramount evidence for this in the morals of prominent sociobiologists themselves, who adhere to a morality that their biological theories cannot authorize but for which they must turn to a liberal Western moral tradition. "The scope of the theory is thus radically limited, since it cannot include the morality and rationality of the authors themselves" (p. 176).

Liberal sociobiologists in fact learn their morality from their cultural traditions, not from their biology, and may even be forced to urge that culture educate within us an ethical sensitivity that transcends biology, rhetorically put as rebelling against our selfish genes. Such ethical sensitivity has had its classical origins in the religious traditions distinctive to the various cultures. This higher morality evidences a possibility not yet accounted for by the biological theory. This possibility has become actual not only in a few sociobiologists who operate "enlightened" (p. 176) with an ethics that exceeds their biology; it has, in fact, been operating in the struggle to be human since such figures as Buddha, Christ, and other classical moral teachers of human history.

What does it mean to say that our beliefs and our ethics evolve no less than our bones and biochemistries, if, as Gilkey holds, this is not the whole story, for neither our beliefs nor our ethics are finally shaped by natural selection, but rather by rational evaluation and by standards of justice? Is the origin of science as much a challenge to explain as the origin of ethics, since neither the truth of a scientific theory nor the rightness of an ethical imperative is to be judged by the numbers of offspring that those who hold it leave in the next generation? Do we need to be free from determination by natural selection to judge whether sociobiology is true, and also to judge right from wrong? Is freedom to judge as important in science as it is in ethics? Where do the sociobiologists get their freedom to evaluate their theories? Where do they get the moral standards they recommend? Do we not need to know the origins of their ethics before we can settle the question of the origin of ethics?

Langdon Gilkey is Professor of Theology, Emeritus, at the University of Chicago Divinity School. He is the author of *Religion and the Scientific Future* (1970), *Maker of Heaven and Earth: A Study of the Christian Doctrine of Creation* (1959), *Creationism on Trial: Evolution and God at Little Rock* (1985), *Naming the Whirlwind: The Renewal of God Language* (1969), *Reaping the Whirlwind: A Christian Interpretation of History* (1976), *Message and Existence: An Introduction to Christian Theology* (1979), and *Society and the Sacred: Towards Theology of Culture in Decline* (1981). During the Second World War he was teaching at Yenching University, near Beijing, China, and was interned by the Japanese at a prison camp near Weihsien, Shantung Province, where he was helper to the camp mason, cook, and kitchen administrator. A theological interpretation of his prison experiences is *Shantung Compound* (1966). He later returned to the East, to Japan, as Fulbright Visiting Professor. He was twice a Guggenheim Fellow and is past president of the American Academy of Religion.

■ ■ ■ ━━━

What can evolutionary biology and theology each say about a subject they have in common, the nature and characteristic behavior of human beings? This is only a part of the larger relation between biology and theology, and trouble along the border between the two is by no means new.[1] Still, tensions today are significantly different from those at the end of the nineteenth century, when theologians attacked the Darwinians. The incursions now come from the other side, especially from sociobiology.

1. *Evolutionary and Historical Human Nature*

Let us begin on a constructive note, namely how much evolutionary biology can tell us all, and so also the theologians, about human beings, about the sources in nature that are determining factors in our becoming and being what we are, and so (note the paradox) how we may become more truly what we are. Sociobiology has begun a concerted effort to do just this. I welcome and encourage it, though I will suggest how it might be done even better!

The reason that biology can tell us so much about ourselves is that we arise from our story, the story of life. In this sense, humans are even more "historical" than those concentrating on cultural history and historical traditions have thought. Our "historical tradition" must be expanded to include evolutionary history. We learn about our formation in a new and deeper way, stretching back almost endlessly to the beginnings of life and even the beginnings of the cosmos itself. What we are, in all the features and aspects of our being, has developed out of that story. In the processes of their development, forms of life have slowly shifted, grown more diverse, more complex, and, in our epoch, resulted in our characteristic identity as humans. All living things, their characteristics and behaviors, come to be over time by a process of slow mutations and the changes resulting from them.

Surprisingly, this is as true of our minds and spirits, our customs and standards, our ideas, our religions, and our intellectual, moral, and spiritual powers as it is of our brains and organs. No more than the psyche or the body (the "soma") is the mind or "soul" or (as we prefer to label it) "spirit" lowered down from above into a material, soulless, mechanical body. Body and spirit together and in strange unity both develop out of evolutionary nature, a point with radical implications not only for the human sciences but also for theology and especially for a philosophy of nature.

Whitehead insisted on the principle that human being in all its outer and inner complexity, its bodily and its mental, moral, and spiritual powers is to be understood in terms of a scientific understanding of nature. But this principle is a two-edged sword: nature edges into mind; mind edges into nature. If the mind must be understood bodily and naturally, nature as the source of human being must also be understood compatibly with those terms through which we understand human being, including mental, moral, and spiritual aspects. In an evolutionary perspective, a reductionist view of nature is only possible if first we hold a reductionist view of human being. This, at bottom, is the dilemma that Michael Ruse struggles to evade (Chapter 4). He reduces ethics to biology, he interprets a distinctively human behavior as nothing but animal genetics, and so the ethical component becomes illusory. As a result, he is unable to understand nature as a precursor of the evolution of the human spirit with its moral life. His nature is incompatible with moral human being.

The influence of this cosmic and evolutionary story, passed on to us through our genes, through DNA, is immense, almost beyond comprehension, and it is not to be denied. We can be grateful for Thomas Cech's discoveries about the origins of life, or for Francisco Ayala's and Elliott Sober's accounts of how human intelligence has been selected for and how conscience has arisen in this process of selection. In any plausible account of human being—which includes all human achievements and behaviors—this historical influence of "nature" (as natural history) must be set down along with the historical influence of "history" (as world history) and the cultural traditions (nation, heritage, community, family, "nurture") as providing the major explanatory factors alike of what we are and do. *All* our possibilities are in the end genetic; our physical, psychological, moral, and spiritual similarities, as well as all our differences, have their patterns set in our genetic endowments. Thus it is true that our common social customs, manners, laws, morals, mores, and our religious rites, myths, and laws, can, or could be, traced back into the dim recesses of the human past. Our behaviors and beliefs evolve no less than our bones and our biochemistries.

Of course, it is also true that the disciplines of biology, medicine, and genetics themselves, in fact the entire corpus of science and philosophy of science, can *also* be traced back there. The *whole* of culture, including science not less than morals, has its *causes* in genetics. I will presently be examining Richard D. Alexander's *The Biology of Moral Systems* (1987). *The Biology of Science*, especially *The Biology of Biology*, is the next book Richard Alexander should write—unless, as I suspect, his theory is only half-hearted. Likewise, Michael Ruse owes us an account of whether his discovery, made from within biology, that ethics is illusory is itself a genetically caused event. Such points aside, this evolutionary development of what we are entails immediately that biological knowledge of the major determining factors in this history will inform us immeasurably about who we are as humans.

It seems evident, on the other hand, that this is not the whole story. Among the powers and the necessities of our human condition, as it has developed historically out of nature, we discover that we are in part self-directing, centered beings. This may possibly be true of some other species of life besides our own. Strangely, we humans must choose, affirm, and use these possibilities of our nature, as we do all other aspects of the cultural, moral, and religious traditions that we inherit. Note that *choose* and *use* are not employed here in their ordinary uses, but in analogical senses. The call to "choose ourselves," to embody in our existence what is given to us, to embody it by inner decision, affirmation, commitment, and perseverance, is one of the characteristics that makes us human; it is the source of responsibility, and so of the moral. It is also, let us note, the ground of deliberation, judgment, and assent; it is also the ground of intellect and of science, morals, and reason. Morals and science alike spring from the *humanum*, the human world—what theology has classically termed "the image of God." This choosing is the referent for the category of "spirit" or "freedom." This requirement to choose ourselves includes the possibilities given us by our genetic inheritance and those given to us by the familial, communal, and cultural inheritances.

Another biologist that we will be analyzing is Richard Dawkins, who, in *The Selfish Gene* (1976), makes strong claims about how human behaviors are genetically based. But before he is finished, Dawkins freely admits this basic power of choice, of self-determining even in the midst of these determining forces that create us. With the causes, there are choices. Our ability to chose between replicating genes or memes (ideas) is what distinguishes humans from animals.

This necessary intervention of autonomy, of self-constitution, is the case with all human existence—whether we speak of the roles of student or teacher, lover or beloved, mother or father, theologian or scientist; none can escape what they are genetically, nor can anyone avoid "choosing" what they are if they would become what they are. We shall presently apply this to the sociobiologists themselves. Human existence is a baffling mixture of being conditioned and of self-determination, of an inherited "given" and a chosen self-direction for the future, of being objectively determined and of living as a centered, subject-spirit. Neither of these apparently opposite poles can be omitted—and both in their mixture arise from nature.

To return to our point about history and to interpret it theologically, this cosmic and evolutionary process is the way that the divine creates such a complex, many-sided, even paradoxical being. We have come to accept such an influence (not to say determination) *historically*, that is, with regard to the influence of cultural history on our own being, because of the power of the inherited historical consciousness. We know that we each are what our social traditions have fashioned us to be, even if also we have chosen these heritages. But this is just as true *naturally*, that is, with regard to the natural history out of which we arise. The ordered set of possibilities, the élan to live among and through these possibilities, and the power or necessity to choose ourselves, which together help to constitute us, are provided genetically by our biological development and inheritance.

Nature has apparently prepared in advance for the various unexpected levels that, in due course of historical evolution, come to make up the whole of nature. We now know as never before that the inorganic contained the possibilities, the ingredients of life—however that emergent came about. Thomas Cech's work on RNA as an enzyme shows a path by which those possibilities were converted into probabilities and then into realities at the first origin of life. This earliest life, single-celled in form, contained the possibilities of more complex mutations. But, as philosophers if not also as scientists, we cannot just assume possibilities. We also have to ask, Where did *they* come from? As Whitehead remarked, possibilities cannot float in from nowhere.[2] These possibilities are the key to the history that we are trying to understand.

Further and more to our point, these expanding complexities in life contained the "seeds" of psyche and of spirit, that is, of purposes, intentions, values, symbols, experiences, fears, and hopes—attempted projects on the one hand and despair at failure on the other. Thus, in effect, all of the richness of culture: art, crafts, myth, morals, politics, religion, science, even universities(!), all of the facets of "spirit" or "reason," the entire *humanum*, stretch back into the dimness and mystery of so-called matter, into the mystery of nature as the source and ground of all that we are. Here nature as "matter"—nature as known by the physical sciences, as *reduced*, one might say—shows its deeper, and more mysterious, identity with nature as *also* the source of psyche and of spirit. Nature as the source of studied *objects*, as objectified, is here united with and enlarged into nature as the ground of inquiring *subjects*, as the source of even the subject-observer. Nature as power and order discloses itself as inclusive of nature as the source of meanings.

Let me draw out the force of this for theology. It is not just our organs and bones that we inherit from the evolutionary process. On the contrary, it is *all* of us, "the whole catastrophe," as Zorba puts it (Kazantzakis, 1953), including morals and religion, as sociobiology says—*and* science, as we remind *them*. Clearly all this developed slowly, at first hidden in other forms of life, then cavorting under various kinds of disguise, of incognitos, then more and more evident, then finally appearing explicitly in ourselves with our consciousness and self-consciousness. There is a compounded awareness: of self and of other; self-awareness of affects, intentions, needs, goals, and values; the power of memory, foresight, and planning; awareness of "world" surrounding and over against us; curiosity, wondering, classifying, theorizing, deliberating and testing—and on and on through all the "human" characteristics that make up art, morals, religion, *and* of course science. Most of these are known to us directly from the "inside" in our own self-awareness and only signalled to us by others through signs and hints on the outside, as we judge from their behaviors and words—a point important in suggesting limits to whether and how far objective, sensory scientific methodology can describe *all* of reality.

These characteristics of "subjects"—that is, of humans, persons, minds, and scientists—are present latently in more subdued forms throughout the process of evolution; they developed there historically. It is, therefore, of tremendous interest to see how they did come into being, what factors brought them about: how needs became values, how needs thus bred signs, as when, to an organism, one thing, a smell, signals another, a predator. We can find out how, in turn, needs bred expressions; the organism does not simply detect but gives out signs, as with courtship displays, and there arose the beginnings of "meanings." Needs, expressed so through signs, consequently bred both hope and despair. Signs passed over into symbols. In time, all of this, the "seeds of spirit," bred customs, morals, communication, intelligence, knowledge, the sciences, and the religions. For *all* these phenomena stretch back into the historical evolutionary past in incognito form, and that "hidden" presence needs to be brought out and articulated.

Classical theology—and with it the rest of Western culture—thought that nature had been created first and on its own, with humans created later and separately. Hence the one, nature, provided us with clues at best about our bodies only but not as to our own deeper nature, what was called our souls. Evolutionary science has taught us how we humans have appeared in all facets of our being *in and through* the processes of nature; hence, a theological understanding of human being must also be informed by a biological understanding. Correspondingly, a historical and theological understanding of all the aspects of culture must be informed by a sociobiological understanding of the biological roots of culture. In turn, of course, and as a consequence, nature must be reinterpreted as the source of culture and so of "spirit." If nature edges into mind, remember also that mind edges into nature.

Such an evolutionary explanation of humans as cultural beings may help us to see how it is, genetically and biologically, that we have been able to become, for example, *scientists*. But such explanations of the genetic origins of science, of the *causes* of science, do not answer the question of whether any science is true or not, or even whether it might be useful. Nor does it answer the question of whether some use of science might be moral. Whether the science of genetics is true is not answered by looking at the genes that genetic scientists inherit and on which they depend as they do their science. An explanation of the *causes* of scientific theories does not answer the questions about the validity of scientific theories, and that goes not only for science but also for anything else. The same is, of course, true of philosophy, morals, and of religion—though I do not think that Alexander, Dawkins, or Ruse have quite seen that point yet. If morality is explained by its genetic causes, and not to be legitimated in any way rationally, so also is the validity of science, and thus Ruse's problem of illusion haunts the scientist as much as the moralist, and both just as much as it does some clockwork machine, where nothing but causes operate. Ayala and Sober, however, are seriously inquiring how our rationality, made possible by our genetic endowments, can also transcend those origins to make intelligence a reliable judge of truth. Perhaps we can find a place for intelligence to make reliable moral judgments. We can next put this to test on the moral judgments that Alexander, Dawkins, and Ruse themselves make.

2. *Ethical Judgments: Sociobiology and the Sociobiologists*

To make this general account of the evolution of human nature more concrete, I shall next ask about the evolution of ethics, both as it originated in the historical past and as, stored in our genes, this origin continues to operate within humans today. We shall have to mix science and conscience, looking both at the science of sociobiology as it makes claims about the origins of ethics and also at the ethics of the sociobiologists themselves. As indicated earlier, I now turn to take up directly two current accounts of sociobiology: Richard D. Alexander's *The Biology of Moral Systems* (1987) and Richard Dawkins' *The Selfish Gene* (1976). Both seek to present on the basis of *biology* a *general* theory of human nature and behavior. I have shown how much theology can learn from the efforts of sociobiology. Still I think these authors need to rethink what they are about in order to be as helpful and creative as they might be. For in the way they are doing it at present, they run into a rather serious set of contradictions.

The dilemmas in these accounts stem from the effort to speak about the whole of human nature and behavior in general in terms of a special, limited, and relatively abstract discipline. Every kind of discipline is an abstraction from the whole of experience. This is something like abstracting one course from the whole of a university curriculum. A discipline thus brings to bear a particular perspective that is almost certainly valid but also partial. Each perspective itself presupposes the whole and thus the other parts of the whole, and it also leaves out much that is important. Hence arise its dilemmas and contradictions when, confined to its own limited terms, it seeks exhaustively to interpret and explain the whole.

Such difficulties are quite evident in these sociobiological accounts: (1) The scientific *subject*, the scientist and so these splendid authors, are omitted in their biological discussions of morality and rationality, or brought in awkwardly at the end of the discussion. Thus, when pressed, these discussions contradict themselves and end in an unexamined dualism. (2) To explain the whole on the basis of a part, words and categories that are defined and intelligible in their limited context must be stretched into rubbery analogies. If unexamined by reflection, these analogies becloud and mystify the theory itself. (3) There is a vast complexity to the human being, to morality and to rationality—illustrated clearly in these two scientific volumes themselves—and such a limited account cannot articulate this complexity.

The most glaring example of such dilemmas can be seen in how both Dawkins and Alexander enunciate, as quite natural and authentic, their own view, which is a very high, liberal, strikingly altruistic set of moral sentiments concerning present-day world problems. Both seem sincerely to recommend "rational science" as a means for resolving these problems (Dawkins, 1976, pp. 2–3, 118, 186, 213–215; Alexander, 1987, pp. 21, 40, 126–129, 182–184, 186, 202–204, 221, 254–256). They do this, however, despite the fact that, according to the biological theory that both recommend, morality arose, together with reason and consciousness, in order through deception and the manipulation of rivals to protect and preserve the self-interested individual organism, which is but the "survival machine" of the selfish gene. Their genetic theory of human behavior spells out a view of human morals and human reason as thoroughly selfish in character—a view quite at odds with the morals and the reason these two authors themselves both recommend and incarnate.

In sum, they have a dim view of the same human reason that they use to commend a high morality, all the while maintaining that this reason typically produces immorality. This dim view of human morality and reason is, incidentally, a view with which any theologian, familiar with the orthodox tradition, feels quite at home. It is, therefore, strange that neither Alexander nor Dawkins, nor Ruse either, seem at all aware of the long tradition of theological reflection on the Fall of Adam and Eve and its baleful consequences, consequences that to everyone else are obvious enough in the overwhelming presence of selfishness, conflict, and suffering in all of human history. For their high view of reason and morality, Dawkins and Alexander consult mostly post-Enlightenment philosophers and their own untutored sense of theologians as naively idealistic and moralistic. While, as we shall indicate, the tradition of Christian theology and current sociobiology are by no means identical, nevertheless there are surely as many formal and material similarities between these two views as there are differences.

Let us look more closely at their main theory, offering a *biological* explanation of human moral behavior, especially of human selfishness. Here we see the deepest similarity between the theological and the biological accounts of selfishness. In each, the evil that we all do is not so much seen as an "act" but as a "*situation*," an unavoidable *condition* of our existence; and in this situation this evil has become a force or power "binding us," often against our conscious wills, to a selfish course of action that we explicitly—though hardly wholeheartedly—deplore. In the biological case, the so-called "selfish gene" programs or rules us; in the theological case, the primordial break in the relationship to God, the "Fall," has distorted all our powers and thus our ability to renew and transform ourselves. In both, therefore, help must come from "outside"—in the one, from objective reason embodied in science, apparently unaffected by the otherwise ingenious gene, and in the other, from the grace of God. Which one of these outside sources of rescue is in fact the more incredible represents a real debate in the twentieth century!

In Dawkins, echoed by Ruse, the selfish gene "creates" its "survival machine"—the individual entity, you and me—for the sole purpose of its own survival, and "programs" this machine (the individual) to preserve the gene's existence over time. Thus it is the gene and its "will to survive" that, as Alexander puts this point, is the "ultimate cause" of all our intentions and acts. All of the individual's desires, emotions, allegedly "altruistic" acts, self-interested acts, and all wayward patterns of behavior are in fact merely surface or proximate effects of that ultimate cause (Dawkins, 1976, Chapters 2, 3, and 4; Alexander, 1987, especially pp. 20, 25–26, 36–40, 64; Ruse, this volume, Chapter 4).

There is indeed a striking similarity to the theology of original sin. The gene's irresistible drive to its own perpetuation affects, infects, and even dominates *all* aspects of the individual person (the survival mechanism): the will, desires, instincts, rationality, morality—and behavior. In classical theology, correspondingly, all aspects of the human being—its will, desires, even its reason and moral standards, and so of course its social existence—are tainted by sin. As Augustine said of Rome and the Romans, their virtues, and their institutions, are merely massive vices [Augustine (c. 425 A.D.) 1950, Bk XV, Ch IV, pp. 482–482]. Alexander, as a biologist claiming to have made an original scientific discovery a millennium and a half later, could well have made the same remark.

In Alexander, echoed again by Ruse, consciousness and reason originate in order to preserve the gene and its survival mechanism in the latter's competition with rivals (Alexander, 1987, pp. 76–78, 100–103, 115–117). Morality, with its conscience, arises both (1) for the reciprocal cooperation that directly or indirectly aids an individual's survival, and (2) in order to deceive others (that one is more rational and moral than one actually is) into being altruistic toward oneself (Alexander, 1987, pp. 102–103, 116, 118–125). Morality has no other function or purpose.

But, meanwhile, share what estimate of human sinfulness sociobiology may with theology, we are left within biology with the problem of obtaining a better morality. Helpful as the Dawkins and Alexander accounts may be in adding a biological account parallel to the theological account, when we try to interpret the deep ambiguity of both moral conscience and objective reason, this theory in biology quite fails to explain (1) how and where these two biological authors derive the high moral conscience and objective reason with which they operate, and (2) how this moral conscience and objective reason might be related, if there *is* any such relation, to the immensely selfish and deceptive morality and reason described in the theory itself. The scope of the theory is thus radically limited, since it cannot include the morality and rationality of the authors themselves. One almost suspects them of envisioning two separate moral species: the ordinary immoral humans of history, driven to unmitigated selfishness and deception, and the other species, the rational and moral scientists, interested only in truth and in serving others. There is a "whiff" here not only of a strange metaphysical dualism of selfish genetics and unselfish scientific intellects, but even more a sociological dualism between lay folk and scientists. When a theologian recalls the theological past, these enlightened scientists have a by no means unfamiliar sense of superiority to the ordinary run of mortals.

But we do not mean to dismiss their high morality as unreal or self-righteous. We have to take it quite seriously; it is a morality that the theologian also may wish to commend. The inquiry we have to press is its origin. Despite their insistence on the sovereign power of the selfish gene over all aspects of human existence, both biologists are witness (perhaps unconsciously) to the claims of a higher morality. This claim has revealed itself *not* in their biological theory—where everything from consciousness to rationality is governed by manipulating selfish interests—but in their *own* splendid moral sentiments, judgments, and hopes for the future. The latter are truly altruistic; universal, just, nondominating, pacifist, rational—goals clearly transcending the nationalistic, class, tribal, sexist, and racist "morals" of ordinary social life.

This higher level of morality (though they never call it that) results from, or is at least incarnate in, science, its knowledge and its objectivity of attitude. But, as we have noticed, there is nothing at all in the science they propose that can ground such a morality, separating off a higher moral species. In fact, they seem to adopt their morality from their and our social inheritance, an enlightenment tradition that we ought to receive and use to reshape our genetically inherited nature. This reformation to morality is not for them "nature"; it is even, as Dawkins admits, "against the genetic inheritance." Consequently, it has to be *taught* to our children (Dawkins, 1976, pp. 150, 214–215). But, of course, it does not answer our question about origins to appeal to an enlightened heritage, now manifest in science. One wonders where this higher morality and objective reason, escaping the influence of the selfish gene, did indeed come from. How did they get enlightened in the Enlightenment? If this morality is not genetic in origin, does it represent a different substance (Descartes)? It could even be supernatural in origin! So far as Dawkins and Alexander are able to give us an account, it is just another one of those possibilities that float in from nowhere. Whitehead warned us to worry about such unexplained possibilities.

The implicit dualism here becomes explicit in the theory of *memes*, Dawkins' term for the units of social inheritance, the more useful and enlightened of which we ought to accept. Memes are ideas produced by a "social federation of brains," bits of consciousness, beliefs capable of evolving, continuing, and flowing down the generations. Memes travel by social inheritance, with important analogies and disanalogies to the way genes travel by biological inheritance. Memes are of many kinds, some better, some worse, but in a growing heritage they create the positive side of culture. The origin of memes is left unspecified, except that they appear as ventured by the mind with its ideas. We have to note carefully that reason is *not* deceptive here. Some memes serve our genetic selfish genes; others do not. We can chose the better memes, and hence we can rebel against the determinism of the genes, against nature.

Thus, Dawkins claims, our ability to choose between replicating genes *or* memes, and also between better and worse memes, is what makes us uniquely human. He reaches (we might even say preaches) a rousing imperative, in the closing paragraphs of his book:

> One unique feature of man, which may or may not have evolved memically, is his capacity for foresight. Selfish genes . . . have no foresight. They are unconscious, blind, replicators. . . . It is possible that yet another unique feature quality of man is a capacity for disinterested altruism. . . . We have the power to defy the selfish genes of our birth, and, if necessary, the selfish memes of our indoctrination. . . . We are built as gene machines and cultured as meme machines, but we have the power to turn against our creators. (Dawkins, 1976, p. 215)

The strong claim of the first paragraph in his book, summary of the first ten chapters, and offered there as "the truth," is: "We are survival machines—robot vehicles blindly programmed to preserve the selfish molecules known as genes" (Dawkins, 1976, p. ix). But that claim is withdrawn in the final chapter, at the conclusion of the book: "For an understanding of the evolution of modern man, we must begin by throwing out the gene as the sole basis of our ideas on evolution" (p. 205). The last word is, "We, alone on earth, can rebel against the tyranny of the selfish replicators" (p. 215). In between, the problem of the origin of such potential is never solved. We need a redeemed reason, motivating us to genuine altruism, but this is only a possibility that floats in from nowhere, leaving contradiction and dilemma between the first page and the last, and leaving us with no adequate explanation of the origin of ethics.

If we are to have a complete theoretical account of human behavior, we will have to know where and how these memes arise, both the better and worse ones, the selfish and unselfish ones, as well as how they now operate. But in Dawkins, one can only wonder *when* this unselfish or altruistic element in the human make-up first appeared. With Galileo or Newton, possibly with Darwin? Somewhere in the Enlightenment? Or, further back, with the origins of the classical religions. With Buddha? With Jesus? Dawkins does think religious ideas are memes (Dawkins, 1976, pp. 207–208). Perhaps even further back. If a *genuine* conscience and a really *objective* reason were there in all of preceding history, say in primordial and archaic cultures as well as in our own contemporary "scientific" culture, then surely such a presence forces a reconsideration of the status and possibilities of "moral systems" in the main body of the theory.

Since the appearance of these "higher levels" in the human self is quite unexplained, not by Dawkins or Alexander, nor either by Ruse, it remains incoherent with, or at least radically disjoined from, the more realistic/cynical biological theory. The entire theory thus represents a kind of unformulated *dualism* made up on the one hand of evolutionary nature set over against rational science on the other hand, a rational science that has no clear account of its own social and historical origins. Perhaps one can interpret the sociobiological account as placing the *object* of the scientific inquirer over against the inquiring *subject*, the scientist himself or herself, to form a sort of up-to-date evolutionary Cartesianism (Alexander, 1987, especially p. 157).

Ayala (Chapter 5) and Sober (Chapter 6) are much more circumspect; neither wants a dualism and both agree with the sociobiologists that the moral sense has an evolutionary origin. At the same time, the human mind, the basis of culture, has striking powers that transcend natural selection. Ayala believes that intelligence was targeted by natural selection, but that when this eminent morality evolves it is a byproduct, as also can be literature, art, science, technology, politics, and religion. Sober believes that the rational and moral processes of culture "float free" from determination by natural selection. We can appreciate here the frankness of a biologist and a philosopher who both refuse to make biology the whole story. But we can also wonder if simply naming the moral sense a byproduct or invoking free-floating mental processes is not floating in possibilities from nowhere, reminding us again of Whitehead's warning. We still need an adequate account of the origin of both rationality and morality.

Meanwhile, Dawkins, Alexander, and Ruse, since they are unable to find any account allowing humans to rise to genuine rationality and morality, have to reduce rationality to something less. Consistent with their view of ordinary, nonscientific reason as fundamentally manipulative, all three develop a complex and interesting theory of *deception*, namely the pervasive deception practiced by animal and by human species, even in the latter's supposedly rational and moral practices. Reason and moral sense are *essentially* deceptive. Since in biology *function* and *inception* are correlated, the one explaining or even "causing" the other, both reason and moral sense arise and so function as mechanisms designed to deceive others; that is, their role is to deceive others into believing the deceiver is more altruistic than he or she in fact is, and this deception is practiced in order to lure victims into becoming themselves altruistic and so into becoming vulnerable to the deceiver's competitive aggression (Alexander, 1987, pp. 100–103, 110–125, 177–178). There is no hint in either text of any principle, genetic or otherwise, that might or could transform or "heal" reason so that it transcends or is redeemed from this essential self-serving, manipulative, and deceptive character. Reason's genetic task is to deceive by all the intellectual weapons it can lay its hand on, and apparently nothing more can be said *biologically* about reason.

But if this is really so, and all there is to be said, the reader cannot help but wonder why and on what grounds, supposing the theory to be valid, the moral sentiments and scientific conclusions expressed by these scientific authors, however persuasive and apparently empirically valid, should not *themselves* be considered to be agents and products of deception, of precisely the intricate, infinitely devious, and yet quite ruthless process of deception described in the theory. Were these books they have written *also* designed and programmed to deceive the reader into some altruistic political act (perhaps to become a believer in genes)? Are these acts of writing just more acts by which the readers are deceived, acts that are somehow helpful to the authors and to their group interests—whatever they may be—in this case, I presume, the community of sociobiologists and *their* genes. Since this application of the theory is absurd, one can only assume that, somehow, to these authors their own moral sentiments and careful rational arguments, being "science," represent a *different* sort of morality and rationality than do those described in their theory. This is not an uncommon judgment about one's own "higher" morals and rationality! But again we encounter signs of a sharp dualism.

Sociobiologists seek to describe, analyze, and explain "morality," that is, moral rules and norms, the distinction between right and wrong, and especially "conscience." All such moral phenomena are interpreted as developments "designed" by the genes to serve the "selfish" purposes of the genes that (who?) have "made" or "created" all the attributes and powers of organisms. So understood, the basis for the so-called "morality" of organisms involves two levels that couple up: (1) most genes will survive and perpetuate only if the whole organism also survives and propagates successfully, and (2) most whole organisms, though individuals, survive and reproduce best in groups that cooperate (with both direct and indirect reciprocity). On this twofold ground, it is essential for gene survival that phenotypes live in groups, that they care for one another (parental, sibling, and kin care), and that they cooperate socially. Hence, from an evolutionary perspective, there arise moral sentiments toward others and the need for moral rules to encourage and even require cooperation. Like civil laws, moral codes are rules of association *for* the survival of the group and were developed largely to express and defend the interests of the group against rival groups.

Translated into the terms of the theory, this means that, in social species, those genes that program their organisms for cooperative group life will be successfully perpetuated and spread. Thus, what is regarded as "moral," even as altruistic, on the part of an individual is, from a biological standpoint, an act that is genetically "selfish." Virtues are really vices. In social life where an individual lives with close relatives, the real "interest" of the gene may frequently be served by the sacrifice (the altruism) of its particular bearer, since other copies of that gene may thereby be given a better chance to survive. Like the deceptive rationality, therefore, morality at the primal level is a reflection of fundamental "self-interest" on the part of the gene, while on the secondary level it is a reflection of the "selfish" survival drives of a given cooperative group in the competition of life. As a consequence of these two factors, there arose historically and still continues to arise parental love, cooperative morality, and even altruism within a given group (Alexander, 1987, pp. 1–3, 37, 72, 77–103, 115–116, 163–164, 177, 182–184; Dawkins, 1976, pp. 7–8, 97–103, 137–138, 179–202; Ruse, Chapter 4). Morality is thus grounded solely in self-interest, at least on the part of its ultimate instigator, the gene; and altruism, while perhaps descriptive of some of the felt "motives" of the individual, is to be seen biologically as a function of genetic self-interest.

This account of morality reminds one of Reinhold Niebuhr's very perceptive analysis of patriotism (1932, pp. 91–93). Niebuhr saw how patriotism combines altruism on the part of the individual with self-interest on the part of the group. Hence, patriotism exhibits a particular "doubled" intensity and fury. The patriot deceives himself into thinking that he is being altruistic, but at a deeper level he is actually being selfish, and so the gains to be had by selfish action are combined with the motivation that comes with an impassioned sense of righteousness. If Dawkins, Alexander, and Ruse are right, all our "moral" actions, so-called, have this doubled, deceived, intensity and fury.

As we have noted, however, this biological description of morality as a function of self-interest—and of all moral idealism as a mode of propaganda or corporate deception—does not explain at all the "high" moral sense of the sociobiological authors themselves, whose prescriptions systematically defy this genetic and group self-love; rather, they urge, as an authentic obligation, a just, fair, universal society based on science and so *not* on deception. They do this with some passion and intensity, but they think they have escaped this deception.

I conclude that this derivation of morality and reason from the "drive" of the gene to preserve itself is illuminating, fascinating, *but*, very speculative. On the one hand, it does make some sense when we look around us at our world, with its regnant self-interest and deception. Sociobiologists are discovering a biological basis for what Christians have long called original sin. But this negative, realistic view must be only one perspective on the fuller truth. Its all-determining selfishness and deception must be qualified so as to enable us, scientists and theologians alike, to explain just as intelligibly the origin of a morality that transcends selfishness and a rationality that transcends deception, of which we have clear examples, let us recall, in the morality and rationality of these authors themselves, to say nothing of the numerous examples commended in the religious traditions. We still need an account of the reformation of human nature, of which these scientists themselves offer an example, even while they are skeptical of the theologians' claims about the origins of morality.

There are three elements in morality as a whole that many commentators on ethics from the sciences overlook. (1) Any account of morality has to explain the *defiance*, on the moral grounds of conscience, of the "moral" customs of groups, just as much as it does the *subservience* of the individual's conscience to the interests of the group. The morality advocated by the sociobiological authors themselves illustrates this point, which they entirely overlook. They defy the inadequate morals they portray in the society they analyze and out of which they themselves have come. Thus the problem for understanding morality in human nature is not only why humans are selfish but also how it is that some occasionally transcend selfishness. On moral grounds, they may risk their existence for an unpopular and rebellious moral cause. (Alexander and Dawkins do, though we could wish for more defiance in Ruse.)

(2) Any interpretation of human being or human morality *as a whole* must move beyond any given science into philosophy. Specifically, it must transcend looking at other groups as *objects*. When we look at other groups, "objectively," as objects we can see clearly their group self-interest, evident not only in their acts but also in their ordinary or customary morality. The reason such an "objective" inquiry into morals is not enough has been revealed in the problem cited above; namely, that a view of morality as a whole must include the moral conscience of the inquirer, the natural or social scientist, whether biologist or anthropologist. For *their* moral judgments frequently (though not always) transcend the moral rules of their society generally, and certainly the rules of any society they look at as observers. Morality seen from the *inside* of the subject, like consciousness seen from the inside as valid cognitive rationality, has a self-transcending, authentic, and responsible character. We take its authenticity for granted, as do Alexander, Dawkins, Ruse, Ayala, Eldredge, Sober, and all other parties to rational debate—and we have to!

For example, in the last century we concluded beyond debate that human slavery is *wrong*. In this century, we are realizing that capital punishment is *wrong*. On critical historical occasions of moral insight, such critical moral integrity has precisely *challenged* and *upset* the reigning morality of the society; these historical instances represent the highest, if not the most common, models of the moral. They are moral breakthroughs, and they may profoundly reform existing morality and shape moral thought thereafter. Historically, Socrates is the great example of this in philosophy, as, in religion, are Confucius, the Buddha, the Hebrew prophets, and Jesus. In our own day, there have been Mahatma Gandhi and Martin Luther King, Jr.

(3) Finally, and most puzzling of all, this critical, universal, and open rational and moral consciousness, evident in all historical life—and very clear in these biological representatives—is itself not as pure as its admirers believe and frequently as its bearers themselves believe about themselves. Even "high" morality—even the morality of liberals(!)—itself illustrates in part the common human condition of bondage to self-interest described so well in their pages. Thus, as the highest religious consciousness has long cautioned, "even the saints know that they are sinners." Awareness of this point is quite absent from these biological accounts of the moral answers to our current problems, accounts that seem

to regard their own resolutions based on scientific rationality and on their own higher liberal morality as exceptions to the problems of genetic self-ishness.

We must, as we earlier said, come to believe that rationality and conscience can somehow transcend the self-interested instincts that we inherit. On this biologists and theologians agree. We are also required, as theologians know all too well, always to recall that this transcendence is imperfect, whether in theologians or scientists. All ethicists remain finite and fallible. Every morality must keep itself in constant self-examination. Such is human nature. In sum, strictly biological accounts of morality remain incomplete because they do not discuss, articulate, or explain the "higher" morality of the scientists and, in consequence, they have no possible reason to become aware of the ambiguity of even that higher morality.

3. *Freedom, Analogy, and Ethics*

One of the main implications of the genetic explanation of evil is its seemingly stark determinism and its corresponding denial of any sort of responsibility—though, again, the sociobiologists exhibit and presumably feel in themselves an inner sense of responsibility for the fate of society and of the world, caught up in the grips of this otherwise inexorable self-ishness. Since, according to the theory, as organic beings we each bear selfish genes that created us for the sole purpose of preserving themselves, we are therefore, as Dawkins puts it, "robot vehicles blindly programmed" (Dawkins, 1976, p. ix), determined to enact these selfish purposes in all we do. Words evoking necessity proliferate in Dawkins' and Alexander's analyses of human nature and are only awkwardly qualified by the sudden advent of the rationality of modern science and its "scientific" morality. Otherwise, this determinism represents an unalterable and sovereign *condition* for all we are and do, the "ultimate cause," as Alexander says, of everything on the surface of life. If the theory is true, clearly there is and can be little hint of responsibility on our part for the fundamental bent of our own "selfish" nature.

Against this, let me suggest, first, that no scientific inquiry can uncover this responsibility and the freedom it implies—both being characteristic of the inquiring *subject* (the scientist) and not of the passive *object* of inquiry. Responsibility and with it freedom are felt *from the inside*, not seen by an observer. Nevertheless, second, the assumption that we are in some way responsible and therefore have some freedom of some sort appears on the fringes, so to speak, of all scientific writing. It appears as the effort to persuade the reader and in the common logical and empirical justification given to each theory—both appealing to an autonomous judgment on the part of a free subject. When invited to criticize any theory, whether in science or in ethics, we must be free to evaluate—actually thinking "inside" ourselves about the theories we are evaluating and the choices that follow, not simply being ourselves blind robots determined by genetic or social forces.

Geneticists never explain genetics by citing the genetic sources (within themselves) of their science. Rather geneticists seek to guide us to *understand* genetics as a valid science, and they do this by citing the empirical evidence and the relevant logical arguments, in short by appeals to validity and not to "genetic" explanation. Each article in this anthology, whatever its content, makes *this* appeal. Thus all parties to the debate—scientists, philosophers, theologians, or whoever—recognize and presume another sort of ground in those to whom they speak and write: the *logical* ground of validity rather than the *causal* ground supplied by the science itself. In biology, this assumption appears in the rational arguments and the responsible, even altruistic, moral stance suffusing most of the biological appeals that touch on the current world problems. "We must use our new knowledge to make the right choices if we are not heedlessly to destroy life on this planet." That is Eldredge's moving appeal (Chapter 2). "We, alone on earth, can rebel against the tyranny of the selfish replicators" (Dawkins, 1976, p. 215). That is Dawkins's last word. Such appeals are pervaded with the language of freedom and so with the awareness, and so the certainty, of our own relative freedom as we face the impinging future. Out of this freedom arises responsibility.

But none of this comes from without, from observing anything, despite Eldredge's descriptions of natural history, or Ayala's description of how conscience evolved as an evolutionary byproduct, or Dawkins's description of blind genetic robots. Each imperative is itself clearly based on self-awareness and in that sense is extrascientific in its character, although it is an intrinsic part of every science that we are obligated to act appropriately in consequence of its findings.

When we locate and recognize this human freedom, we can better see how biology joined to ethics risks the misuse of language, making an inappropriate use of analogies to extrapolate misleadingly from nature to human nature and vice versa. As we noted at the outset, problems of language abound when a particular language appropriate in a limited context is used to interpret the whole. Correspondingly, such problems are very evident in biological accounts of this genetic inheritance as the "ultimate cause" of human behavior as a whole. Sociobiologists find analogical language unavoidable; their use of it appears on several different yet quite undiscriminated levels. This lack of discrimination results in little capacity to deal with the real, experienced sense of human freedom.

First, there is the universal use of the language of purpose in biology generally, for example, "in order to" and "such and such is developed for this or that function." In modern biology, of course, this purposive or teleological language must, despite its ubiquity, be translated into the nonpurposive language of the mechanics of natural selection. But that means, in the analogical extrapolation, that the freedom to act purposively is lost in a mechanics of causation. Second, analogies appear with the shift of dimension or level from *conscious* purposes in human existence, where this sort of language originated and is mostly used, to describe and articulate *unconscious* yet apparently purposive behavior in animal, plant, and organic life generally. These are the only levels of analogy of which Dawkins and Alexander seem to be aware (Dawkins, 1976, pp. 50–53, 132, 149, 156–157).

Third, this is not the most important instance of analogy in these works. Most interesting of all, analogy appears with the shift from the conscious *and* the unconscious "purposes" of the organism or phenotype ("survival machine") on the one hand to the vastly different level of the gene itself on the other, an *immense* shift (Dawkins, 1976; p. 50; Alexander, 1987, pp. 14–19, 34–40). The gene is not an organism, but the "maker" of organisms; consequently, it seems dubious whether it *has* a drive at all. If it does, this is surely a very different sort of "drive" than an organism might have. It is some sort of internal élan or principle unfolding over time, pressing toward its own preservation or fulfillment. Just what it means analogically to say that genes have purposes and can behave selfishly is not clear at all. Yet such language is continually used without reflection by both Dawkins and Alexander. They deny, of course, that in the genes these are *conscious* purposes as they are with us (Alexander, 1987, p. 19; Dawkins, 1976, pp. 24–26, 50–54, 117, 132), but that is hardly the point. The question is whether language about a drive, an internal purpose, or an intention has here any referent at all, and what sort of referent that might be—the questions that arise with any sort of analogy. Where is the analog, if there is one, to that sense of freedom, felt from "inside," when we choose our purposes and feel our responsibilities? If there is none, perhaps the analogy fails. Without something akin to what Aristotle called an *enteleche*, or Bergson an élan, without a "will to survive," or some "self" to be selfish about, without some identity and an impulse to its conservation—something that we can find present in the case of animals and plants—it is hard to know what all this speech *means*—a problem uncomfortably ever-present in theology!

Meanwhile, the use of purposive language for the gene pervades sociobiology, especially Dawkins, Alexander, and Ruse, although Dawkins' usage is, so to speak, "fruitier." "The gene is the most fundamental unit of self-interest" (Dawkins, 1976, p. 12, p. 39); "its business is . . . ," "its task is . . ." (p. 40); "what it is trying to do is . . ." (p. 95); "it manipulates for its own ends" (pp. 36, 47); "the true purpose of the gene is to survive" (p. 47). "Genes are responsible for their survival in the future" (p. 24); "for its own selfish ends" (p. 46); "they (genes) are programming for their lives" (p. 67)—and on and on. The point is not whether or not the referents of this purposive language (the genes) are, like ourselves, conscious. The point is whether the referents of this purposive language have any "intent" of *any* sort, even of the kind that an animal or any organism surely does.

So is there any sense of *any* sort to all these analogies? Is there any drive, élan, or intent in the gene that can be the referent of the analogy of "selfishness," as there is in the life of the organism or animal, or as Aristotle believed there was when he spoke of the *enteleche* or *formal cause* of development in what he called vegetable and animal nature? Without an analogical referent of *some* sort, it is hard to know whether we are talking about anything at all. Is it, then, purely *speculative*, an inference that there must be a drive, intent, or purpose of some sort in the gene? None of the felt sense of purpose or of responsibility really carries over to the gene, even though the gene is covered over with language that suggests purposes and intentions, unworthy selfishness, and bad morals. Meanwhile, the gene, lamented for its selfishness, isn't free to do anything at all, in the moral sense of freedom by which humans evaluate and choose to follow moral claims. Do we have any idea what we mean here? Surely some discussion of language and of analogy—a philosophical discussion—is necessary.

As a result of these three distinct levels of analogy (gene, organism, moral agent) and their mingling without discrimination, the linguistic confusion is immense. Analogies such as "mechanism," "programming," "blindly determined," on the one hand, and analogies such as "choices," "decisions," "intentions," and "rebellions," on the other hand, appear on the same page and in profuse reference to the same purported entities (Dawkins, 1976, p. 132).

4. *Experienced Wholes, Abstracted Parts, and Scientific Mythology*

In sum, the main problem of scientific accounts of human nature as a whole is that they seek to interpret the *whole* of experience from the vantage point of an *abstracted portion* of it, whether it be a perspective that is chemical, physiological, biological, neurological, psychological, economic, political, *or* theological. What is needed is the mediation of coherent *philosophical reflection* to generalize and integrate the partial abstractions of these special sciences to enable them to include (1) the rational and moral *subject* as well as the object of inquiry, (2) the *whole width* of experience—personal, intellectual, social, and historical (and so historically scientific as well as empirically scientific), and (3) thus to develop categories, language, and analogies that can deal with the comprehensive width and strange depth of experience. Without this philosophical mediation, reflection is left with one of two difficult alternatives: (1) *reductionism* of the whole to one abstract area within the whole, such as to physics or biology—a reduction that leaves out the scientific author, or (2) an incoherent *dualism* in which the personal, rational, and moral subject (the scientist) is arbitrarily introduced into a theory developed from inquiry into objects alone. The sociobiologists we have examined have struggled with both alternatives, further troubled because unexamined analogies confuse and becloud what is said. They fail to realize how, in either alternative, since the discussion purports to be about the *whole* of human experience and to present a theory about human nature *as a whole*, it must include the *subject*, intellectual and moral, as well as the *objects* of inquiry. They fail to include themselves, though they too are, inescapably, part of the whole.

One result is that the conclusions of such a discussion and the theories that result from it, since they now include much more than the restricted area of any special science, do not any longer bear the *authority* of that special science. The conclusions from the special sciences surely bring with them great authority when they enter such a general discussion about human being; but there they must join *other* perspectives—for example, interpretations from other physical sciences, from psychology, social science, anthropology, religion, morals, and law. Those disciplines too bring no less authority. All of these deserve a hearing when we try to fill in the entire picture, including the subject, culture, morals, and rationality, and so of science itself. In relation to the whole, each remains an abstraction, a limited field. The conclusions from such a special field, stretched (by unexamined metaphors and analogies) to take in all of experience, remain at best speculative—certainly until they have been subjected to the criteria appropriate for philosophy, namely coherence with other forms of knowledge and adequacy to the entire width of experience. Any conclusions about biology, ethics, and human nature will require an interdisciplinary synthesis.

Stephen Toulmin terms precisely this move from a *special* discipline to articulate the *whole* of experience a "scientific mythology," something that, though it may be launched in a science, is no longer science at all (Toulmin, 1982). A "scientific myth" occurs whenever a scientific category, concept, or formula, which has been carefully defined, tested, and refined within a limited context, is taken out of that context, extrapolated, and used to explain ranges of data, puzzles, and questions far beyond the scope of that limited context. Toulmin cites the use of the category "evolution" as the explanatory category for the entire cosmic process. One can add the Marxist material dialectic as a category from economic analysis used to explain the whole course of history. Such use of categories from special disciplines can and has often become immensely creative, *if* this move to another range of explanation is recognized—and if it is recognized that one has then entered the realm of philosophical discourse and must therefore submit to the responsibilities and the criteria of philosophy.

This limit is as true for theology as it is for the special sciences. Without philosophical mediation, theology can become merely mythological, even superstitious. Biological theory can, as sociobiologists show, certainly contribute greatly to the understanding of human nature; sometimes it can join with theology in insights about human selfishness. But in reductionist and dualist forms such theory does not bear the authority of biological science; rather, linguistically, it functions as myth, and as incoherent myth at that. It awaits the *synthesis* of its findings with those of other perspectives on the mystery of the human being. Such a synthesis is the classical enterprise of philosophy, the sort of "faculty club of the disciplines" where disciplines can talk together, each be elevated and also disciplined by the other, and so reach a common understanding—or at *least* achieve a shared discourse.

Each special perspective tends to see itself as providing *the* key that will unlock the mystery. Theology certainly has made this claim and did so successfully for many generations. So in our day have economic theory (in Marxism) and psychology (in psychoanalytic theory and in behaviorism); and now we see it with evolutionary biology. Perhaps history also shows that such claims are not finally antithetical to the creative advance of understanding; in fact, we can most effectively further self-understanding through testing the special perspectives to see whether they can be enlarged, whether they have a fitness for comprehension of the whole. That is why a colloquium of the disciplines, as represented in this anthology, is of critical importance.

These claims by each discipline, if untempered and unqualified by such interdisciplinary reflection, also illustrate the "shadier" side of rationality, its tendency, however objectively "academic," to see itself as *the* central discipline essential to all the others, to see itself as *the* key to the truth all others seek. Each discipline has an implicit imperialist urge lying in wait for the chance to advance its claims on the others—as we theologians well know! Here, ironically, the influence of the selfish gene seems to have penetrated even the sanctuary of scientific genetics and to have urged geneticists to make extraordinary claims for their own discipline. Though only a part, sociobiology "selfishly" claims the whole! The ambiguity as well as the creativity of reason, even of so-called objective, scientific reason, thus reveals itself even within this purportedly objective discussion of selfishness. Again, what is needed is not only brilliant, articulate defense of what one knows in one's own discipline, but also humility about the limited character of what one knows and about its cooperative place in the entire panorama of human understanding. What is called for is a *synthesis* of all these perspectives, each of which sheds its own light on the larger mystery.

References

Alexander, Richard D. 1987. *The Biology of Moral Systems*. New York: Aldine de Gruyter.

Augustine, Bishop of Hippo. [c. 425 A.D.]1950. *The City of God*. New York: Modern Library.

Dawkins, Richard. 1976. *The Selfish Gene*. New York: Oxford University Press.

Kazantzakis, Nikos. 1953. *Zorba the Greek*. New York: Simon and Schuster.

Niebuhr, Reinhold. 1932. *Moral Man and Immoral Society*. New York: Charles Scribner's Sons.

Toulmin, Stephen E. 1982. "Current Scientific Mythology." In *The Return to Cosmology: Postmodern Science and the Theology of Nature* (pp. 21–32). Berkeley: University of California Press.

Whitehead, Alfred North. [1929]1978. *Process and Reality: Corrected Edition*. New York: Free Press.

Endnotes

1. Elliott Sober, as a philosopher of science, notices that there are frequent border disputes even within the sciences—biology versus physics, social science versus biology—and that in the history of science there is no particular reason to favor resolution in favor of the "harder" discipline (Sober, Chapter 6).

2. "Everything must be somewhere. . . . Accordingly the general potentiality of the universe must be somewhere. . . . It is a contradiction in terms to assume that some explanatory fact can float into the actual world out of nonentity" (Whitehead, [1929]1978, p. 46).

Darwinism and Postmodern Theism

8

Darwinism and Postmodern Theism

CHARLES BIRCH

■ ■ ■ *Editor's Introduction*

We heard Edgar Mitchell recall Earth's "rising gradually like a small pearl in a thick sea of black mystery." Such experience produces a respect, even a reverence for Earth. Sagan and Margulis, facing Earth, see Gaia. Mitchell himself continued, "My view of our planet was a glimpse of divinity" (quoted in Kelly, 1988, at photograph 52). For Charles Birch, too, an ecosystems biologist renowned for his studies of the distribution and abundance of animals, his view of the planet is a glimpse of divinity.

Gilkey, the professional theologian here, chose to emphasize the *humanum*, the human world, and to insist that humans reached levels of morality unauthorized by the current too-narrow models in biology. That *humanum* touches what theologians see as the *imago Dei*, the image of God, in which humans are said to be created. After that, Gilkey chose to leave his theological claims more implicit. Birch, in this respect alone among the biologists in our volume, is more explicit about religion. He too complains of trying to reduce biology to something less than it fully is. The more adequate approach is the other way around, to embed biology in a comprehensive metaphysics. That is not necessary for laboratory work, but it is necessary if a biologist is to be philosophically competent. For Birch, this metaphysics is process philosophy and theology.

Thomas Cech found life intrinsic to the earthen molecules. Birch will say "Amen!" to that, and then add: Yes, but Cech's story is only about the launching of life. Now, at the end of this volume, it is time for a more finished, comprehensive view. When the fuller story is told—how in this planetary setting life rose from the primordial seas, persisted over the geological epochs, and became reflectively self-conscious in humans, who have struggled through the millennia of civilization to understand who and where they are, what they ought do—all of this story must be intrinsic in the molecules. The whole "process" is possibilities becoming actual, and this requires that the possibilities be there at the start—or else they could not be realized in the end.

But we need an account of how these possibilities become actual. On the one hand, those possibilities do have to be there from the start. That is why Birch rejects mechanism. "The aboriginal stuff, or material, from which a materialistic philosophy starts is incapable of evolution" (p. 207). Matter is not inert and passive, mere matter, as though it were like billiard balls, only acted on from without; matter is restless, creative, prolific. That much is attested already by the replicating RNA molecules, self-assembling themselves into life, discovering and storing information about how to make a way though the world. Neither is active matter mere object, "dumb" (like billiard balls); there is subjectivity present, attested in the later evolution when consciousness arises, not *ex nihilo*, from nothing, but by the processive deepening of what was already foreshadowed there, "subjectivity." "Evolution is not simply the evolution of objects. It is the evolution of subjects" (p. 208).

To explain this long story of life wending its way through the tumultuous historical contingencies, to explain the genuine discoveries by which life climbs from protozoans to people, we need not only the possibilities there from the start, but a lure for them. It is not enough to have only intrinsic possibilities; there needs to be a force able to educate these possibilities into actuality. That lure is God. To the "bottom-up" atoms and molecules self-assembling themselves, there must be added a "top-down" explanation that superintends the whole. There must be a principle of orchestration, and more, an Orchestrator (p. 205). That is the divine eros, glimpsed with an overview of planet Earth—not Gaia, but God.

The conclusion to draw is neither that all is determined, nor that all is chance, but that there is a principle of order using chance creatively. That Organizer is the divine lure, and the postmodern model for creation is influence, "persuasion," not mechanistic manipulation, "tyranny." The supernaturalist God belongs to a bygone era in both physics and biology. Birch dislikes Newtonian biology, biologists with physics-envy. Such envy lets physics trump biology. Sober has already warned that, when there are border disputes, the presumption that physics always wins need not be so, nor that biology always trumps the social sciences.

Gilkey has extended this principle to warn that neither does biology trump the humanities. Birch insists that we should not let biology prohibit rising to religious world views, any more than we should let biology be reduced to physics. Biology is only one discipline among others and it must take its appropriate place in the larger story. That larger story takes this biologist, at least, into theology.

Is it a complete explanation of an event to discover that it is natural? Do we need to suppose some "lure" in natural history by which we get steadily more out of less? Otherwise, do we have possibilities floating in from nowhere? Something coming from nothing? Is the mixture or order and chance that we seem to find in natural history godly or ungodly? Is Birch right when he holds that the "by chance" explanation is especially unsatisfactory at the emergence of the first life and at the first emergence of humans with their mental and moral life? What does it mean to say that if "nature edges into mind; mind also edges into nature"? (p. 197) How does the role of chance make possible life, evolutionary natural history, and ethics (p. 209), even though "there must be something positive limiting chance" (p. 205)? Does the the dramatic evolution from protons to people call for religious explanation, with all lesser explantions found to be inadequate?

Reference

Kelly, Kevin W. 1988. *The Home Planet.* Reading, MA: Addison-Wesley.

Charles Birch is Professor of Biology, Emeritus, at the University of Sydney. He is the author of (with H. G. Andrewartha) *The Distribution and Abundance of Animals* (1954), (with H. G. Andrewartha) *The Ecological Web and the Distribution and Abundance of Animals* (1984), (with John Cobb) *The Liberation of Life: From Cell to Community* (1981), *Nature and God* (1965), (with P. Abrecht) *Genetics and the Quality of Life* (1975), and *On Purpose* (1990). In addition, he has 150 published articles. He is a member of the Australian Academy of Science, a fellow of the American Association for the Advancement of Science, and received the Eminent Ecologist Award of the Ecological Society of America. In 1990 he received the Templeton Prize for Progress in Religion, often said to be "the Nobel prize in religion."

I like to ask graduate students, "Need a scientist's philosophical linen be as clean as his laboratory glassware?" There follows a deathly silence! The answer, I suggest, depends upon the matter under consideration. If we are concerned about what experiments the scientist is doing with his glassware, like those Thomas Cech does in his laboratories on the role of RNA as an enzyme, the answer seems to be no. Or, as one scientist said, philosophy of science is as relevant to what one does in the laboratory as ornithology is to watching birds. When you read a scientific paper you don't ask, "What is the author's philosophy?" You ask, "Is this good science?"

I put this question to a great exponent of the philosophy of science, Sir Peter Medawar: "Does your understanding of the philosophy of science influence the experiments you do in your laboratory?" He replied immediately, "No." Then I asked why he was so concerned about writing books on the subject. He replied, "So that I don't make a fool of myself when I relate my science to broader questions in society." Scientists often do not worry much about their philosophical linen; still, since philosophy is the love of wisdom, we biologists do not want to be found fools on the broader philosophical issues.

My topic concerns a strictly scientific theory, namely Darwinism, a theory that underruns each of the essays in this anthology. The other authors have been trying to relate that theory to broader topics, to the evolution of culture and society, to the evolution of ethics, to what we ought to do. Now I want to relate it to a still broader topic, namely theism. That might prove foolish, but, then again, we are not going to be wise until we have the whole picture, a philosophy suitable for our science. How I or anyone else might relate these two depends very much upon one's philosophical world view. The diversity of views on the subject is due mainly to the diversity of philosophical world views. So we need to examine not just the Darwinian scientific theory but, even more, the Darwinian world view that has been founded upon it. We will also find that the world view partially precedes the science; the Darwinian theory is the result of a world view that came before it and makes the theory possible. At this point, washing our philosophical linen is important if we are to come clean about what our science means.

1. *Modern World Views*

It has been said that the best measure of a scientist's influence is how long he or she can hold up progress in his or her own discipline. I shall argue that biologists who are wedded exclusively to the world view of mechanism have held up progress in solving some of the more subtle problems of evolution. Those subtle problems especially include the events surrounding the origin of life and the events surrounding the origin of human life, where a reduction to mechanism holds up progress in understanding. Understanding these events both biologically and philosophically is going to require a more comprehensive view. My argument requires, first, saying something about world views.[1]

The dominant scientific world view is the philosophy of mechanism that became established in the seventeenth century through the influence of persons such as Galileo, Mersenne, Descartes, Boyle, and Newton. Each of them agreed that nature is composed of things that can be called substances, which are devoid of self-motion. Each thing is moved by other things, by what can be called external causes. The natural world is matter in motion, inert things that are externally moved, as when a rock is given a push. Things are not moved by aims or purposes. They have no internal principles of unrest. They just stay around until some other thing moves them. Every present state of a thing is determined by something else, which in turn is determined by something else, and so on back. This is complete determinism; it makes the prediction of each and all events possible in principle. The astronomer Laplace said, if he could know the position and momentum of every particle in the universe he could predict the future of the universe completely.

This doctrine of inertia works well with steel balls sliding down inclined planes and with stellar masses in orbits, the objects to which it was first applied. But what about living things? Did these founders of the modern scientific world view believe that everything, including themselves, could be understood in purely mechanistic, deterministic terms? What about the origin of living things? Is that more mechanics, mere mechanics? When Thomas Cech discovers how chemical precursor molecules assembled themselves into fragments of RNA, which began to catalyze themselves and assemble together into longer chains that could discover and store information and perform metabolic procsseses, nature does not seem so inert and passive. It seems much more creative, as he fully recognizes. Perhaps there is more to the universe than mere matter in motion?

What about human life? Was the movement of Newton's mind, through which he discovered the laws of motion, explicable on the same terms as the movement of steel balls on inclined planes? Or is rationality more than mechanics? What about ethical obligations? Every contributor to this volume has concluded that humans have moral duties in the world. Are those duties a kind of matter in motion? Matter edges into mind, but, as Langdon Gilkey reminds us (Chapter 7), we have also to ask whether mind edges into matter. Matter is what it does, and if matter makes mind, what then?

That sort of question brings us to a great divide in modern thinking. There are really two versions of this modern scientific world view: a supernaturalistic, dualistic version and an atheistic, materialistic version. The former was Descartes' view; the latter is now dominant. An examination of the tension between them will help us see how biologists are helped and also hindered by the condition of their philosophical linen. Both versions agree that the fundamental units of nature are bits of stuff wholly devoid of self-determination or self-motion. But they disagree on whether reality as a whole is composed entirely of such things. A thoroughgoing materialism holds that there is nothing but matter in motion, and that this includes human life as well. But Descartes drew the line between the human mind and the rest of the world. Dogs are barking machines moved only by other things. The motion of all things other than humans has an external cause; it is loco-motion, meaning the motion from one locus to another.

Humans are different. Though their bodies are matter in motion, they have minds as well. They are self-moved. The motion of self-moving things is internal; it is internal becoming. It is motivated. Descartes said that he can move himself from this place to that in space or in his thinking because he has an inner reality, his mind, which is different from inert things. The mind has some degree of self-determination. Descartes said, "I *feel*, therefore I am." For him the most real thing about him was that he had feelings, something that other animals and inert things do not have. Indeed, for Descartes nothing else but humans has this inwardness.

We can begin to see how this world view made biology difficult; the biologists seem to have little choice but to reduce biology to physics, except for retaining felt emotion and motivation in human life. The origin of life, and all of subsequent natural history, was an affair of matter in motion. By this model of nature, evolutionary life was inert matter getting pushed around. Only the human inner life, rationality and our felt experiences, including those of moral obligation, still have a place, and there biology is caught in a dualism of matter and mind. We can almost sympathize with those who were tempted to reduce biology to mechanism, even though they have, unfortunately, held up progress in the discipline of biology.

Seemingly, the only other choice was supernaturalistic theism, the version of mechanism that Descartes held. Not only is there human mind, where mind moves our material human bodies, but also there is God, who moves material bodies. God comes into the dualistic picture because God put matter, which is otherwise inert, in motion at the creation of the universe. God is called the first mover, which is the main description of God in modern discourse. So all power of initiating motion was restricted to God and to self-moving created things, namely humans. This supernaturalist/naturalistic account was the dominant view among those who first formulated the mechanistic view of nature from Galileo to Newton and Descartes. Descartes, we ought to remember, was an engineer before he was a philosopher. Descartes constructed an engineer's universe; and, accordingly, his God is a divine engineer. This engineered world was the view held by Darwin when he was formulating his account of creation.

But the supernaturalistic view has gradually waned, owing in large part to the rise of Darwinian biology and the problems it presented. By the second half of the eighteenth century, this dualistic view became transformed into scientific materialism. By the end of the nineteenth century, it had become dominant throughout the scientific world. There were many reasons why supernaturalistic dualism was so unsatisfactory. I can best illustrate that by the problems it was to raise for Darwin. Darwin needed not only a first mover to start up the world of physics, but he needed to explain evolutionary motion (if we may phrase it that way). He needed to explain the origins and the continuing history of life, biology from its first start on through the evolution of human life. Supernaturalistic, deistic, dualistic theism couldn't help him do that.

2. *Darwin and Supernaturalistic Theism*

Darwin began his voyage around the world a convinced theist. He had read Paley's *Natural Theology* as a student at Cambridge University and was impressed by its arguments for the existence of God from the design of nature. The doctrine of divine carpentry, as it has been called by a later vice-chancellor of Cambridge, was promulgated by bishops from their pulpits. Students at great universities were expected to believe it. Scientists were expected to provide more and more evidence for it. In that respect, Darwin became a traitor to the cause. His voyage around the world was completely to change his view of the source of the order of nature. Supernaturalistic theism had a number of problems for Darwin.

1. Instead of the world and all its inhabitants having been made out of nothing through instant acts of creation, Darwin showed that nature was not made complete and perfect once and for all time. Nature had always been, and still is, in the process of being made. There never was any once-upon-a-time, grown-up creation; if there ever was creation, it was a long-continuing, slow, evolutionary creation. Darwin was not yet in a position to speculate about the origins of life, though this too we are now understanding as an incremental process; but he was in a position to know that the many life forms had not all been created at once, but had evolved slowly, out of simpler forms, over long periods of time.

2. But it was not just the difference in temporal rate, instantaneous versus incremental, that troubled Darwin. Instead of nature being benign, evolution involved a struggle for existence. Most of the creatures that are born into the world never reach maturity; they are eaten, die of diseases, or starve. Even before Darwin, Tennyson recoiled from the savagery in wild nature, calling it "nature, red in tooth and claw" (Tennyson, 1850), and Darwin's theory greatly emphasized this struggle for survival.

Darwin felt that this struggle was ungodly: "I cannot persuade myself that a beneficent and omnipotent God would have designedly created the Ichneumonidae with the express intention of their feeding within the living bodies of Caterpillars, or that a cat should play with mice. Not believing this, I see no necessity in the belief that the eye was expressly designed" (Darwin, 1888, 2:312).

3. Evolution by natural selection involved chance. If random variations are selected for their fitness, this is blind and accidental. That seemed in opposition to the concept of an all powerful designing creator. Darwin wrote before our modern understanding of how variations arise in genetic mutations, but he realized that the variations were arising by chance. So he had to attribute a role to chance in evolution, but he was unable comfortably to accommodate this role of chance in his thinking, either in his science or in his theology. Even scientifically speaking, in principle, he could not admit the reality of chance. In this respect, he was

like Einstein a half-century later, who, when faced with a seemingly random element in physics, insisted, "I, at any rate, am convinced that *He* [God] is not playing at dice" (Born, 1971, p. 91).

Darwin must have greatly admired the deterministic universe of Newton and the sort of thinking that led Newton to that concept. Darwin's dilemma is poignantly expressed in a letter to the Harvard botanist Asa Gray in 1860: "I cannot think that the world is the result of chance; and yet I cannot look at each separate thing as the result of Design. . . . I am, and shall ever remain, in a hopeless muddle" (Darwin, 1888, 2:353–354). "But I know that I am in the same sort of muddle as all the world seems to be in with respect to free will, yet with everything supposed to have been foreseen or pre-ordained" (2:378). Again and again his letters reiterate the refrain, is it all ordained, or is it all a result of chance? Darwin actually suggested that perhaps the solution is "designed laws" of nature, with all details, good and bad, depending upon "what we call chance," as Charles Hartshorne has pointed out (Hartshorne, 1962, p. 207). But that hardly seemed satisfactory either. So Darwin was at a loss how to fit chance into either his scientific determinism or his theological ordination. "The more I think," he lamented, "the more bewildered I become" (Darwin, 1888, 2:312).

Why? Hartshorne has two suggestions (Hartshorne, 1962, p. 207):

1. Darwin tended, like so many others, to think that science is committed to determinism. What we call chance may not be real chance at all. The word *chance* does not always refer to the absence of any determining causes (complete chance), but typically refers to human ignorance of the causes. An event is fully caused, but we do not know what those causes were, so we just put it down to chance. Darwin wanted to stick to a completely deterministic world of nature. He rather liked the notion that God may have started the whole thing going, together with a set of deterministic laws of nature the universe obeyed, and then left it to itself. The universe ran in clockwork fashion, without real chance, but many events, from the point of view of humans with very limited knowledge, looked as though they happened by chance.

The role of chance in our present understanding of Darwinian evolution is quite specific and different from either complete randomness or mere human ignorance of causes. There is a way in which a thoroughgoing determinism can also produce chance events. A chance or random mutation can be one that is fully determined by antecedent events; nevertheless, the mutation is *not* designed to produce a change that would necessarily be advantageous to the organism at the time. The mutations take place at the molecular, genetic level, as a result of causes, among which are chemical mutagenic substances and radioactivity. But events in the causal chains at this level are independent of the level at which biology operates as the organism functions in its ecosystem. The mutations are caused, but blind to the needs of the organism so far as its metabolism, morphology, and the ecosystem niche that it occupies are concerned.

Indeed most mutations are disadvantageous. Mutations occur in great diversity all the time, but whether any of them happen to be of advantage to their possessors is a matter of chance, in the relative sense that the mutation is produced in a causal line that is unrelated to the causal line at which the advantage occurs (relative chance). We say that, in an insect, the mutation of a gene to produce resistance to the insecticide DDT is a chance event. Whether or not a particular mutation will increase the chance of its possessor to survive and reproduce is dependent upon a second chain of events quite independent of the first chain, that of mutation itself. The second chain of events has to do with the environment in which the organism is living at the time as, for example, whether or not its environment contains DDT. Mutations that might have conferred resistance to the insecticide DDT in insects presumably were occurring long before humans invented this chemical. But these mutations gave no advantage to the insect until it found itself in an environment with DDT present. Before the invention of DDT, there would have been no selection of such mutations; afterward, these mutations are selected in the lucky insects.

In this sense, the word *chance* does not imply without a cause; rather, it means that the intersection of two causal pathways "is not decided by any agent and is not fully determined by the past" (Hartshorne, 1984, p. 16). Most "accidents," as we normally call them, have "causes." So when a jet plane crashes, we call it an accident, and yet we look for the causes, perhaps a critical engine part that failed due to metal fatigue. We mean that nobody planned the accident, and that the causal chains that led to the motor failure intersected with the causal chains of passengers flying on the jet plane, and that such an intersection, and tragedy, is not determined by the character of either of the causal trajectories. It is this sense of chance, a lucky intersection, that present Darwinian theory requires, two unrelated causal chains intersect, with the fate of the organism hanging in the result. By this account, Darwinians can have a deterministic universe, yet also one that generates chance mutations.

2. Not only did Darwin want a deterministic universe in his science, but, further, it was not apparent to Darwin why cosmic purpose should leave anything to real chance. God was identified with absolute law and non-chance, the God who created the deterministic universe. The supernaturalist theology of Darwin's day was of no help to him in this respect. God must do everything or nothing. And if God is responsible for everything, then why all the evil in the world? Darwin was left at an impasse. There seemed no alternative between chance and God. Faced with such alternatives, in the subsequent debate, contemporary biologists have come to emphasize even more the chance element in natural history, and this has led Jacques Monod, Stephen Jay Gould, and Richard Dawkins to reject theism.

Only one of Darwin's correspondents, the English vicar and novelist Charles Kingsley, ventured the idea that God could be other than an omnipotent determiner of all the details of nature. Kingsley wrote to Darwin, "I have gradually learnt to see that it is just as noble a concept of Deity to believe that He created primal forms capable of self development into all forms needful . . . as to believe that He required a fresh act of intervention to supply the lacunae which He Himself made" (Darwin, 1888, 2:288). In *The Water Babies*, an evolutionary epic that Kingsley wrote for his children only four years after publication of the *Origin of Species*, he tells how God "makes things make themselves" (Kingsley, 1930, p. 248). There is no evidence that Darwin appreciated Kingsley's attempt to rescue him from the concept of the omnipotent, deterministic God of supernaturalistic theism by pointing to creativity in the creatures themselves.

Darwin is commonly supposed to have shown that there is no room for God or purpose in nature. Herbert Spencer (recalled in Chapter 2 by Sagan and Margulis) feared that ethics was impossible because Darwin had disproved God. Sir Gavin de Beer concludes in the (supposedly authoritative) *Encyclopedia Britannica* that "evolution is accepted by all biologists," and that "Darwin did two things: he showed that evolution was a fact contradicting scriptural legends of creation and that its cause, natural selection, was automatic, with no room for divine guidance or design. Furthermore, if there had been design it must have been very maleficent to cause all the suffering and pain that befall animals and men" (de Beer, 1973–74, 7:23). But Darwin never showed anything of the sort. What he did show was that existing concepts of design by an external agent were invalid. John Passmore, in a more careful analysis than de Beer's of how philosophy and biology mixed in Darwin, says, "Darwin's theory did nothing to prove that God did not exist; but it did destroy the only argument, many men thought, by which God's existence could possibly be established. That is why it encouraged the envelopment of agnosticism" (Passmore, 1959, p. 14).

This element of chance and whether it can be fitted into some alternative theistic picture remains the principal issue. How can God order a universe with chance in it? At the same time, even from a scientific viewpoint, chance, and nothing more, is not an entirely satisfactory explanation, because, after all, there has been a long-standing generation of order over the epochs of evolutionary natural history. Chance can seem an especially unpromising explanation at the two critical events of particular concern, the origin of life with which biology begins, and the origin of human life with its rationality and moral agency. Thomas Cech's discovery of how RNA acts as an enzyme, for instance, dramatically increases the probabilities of life's origin; in his view, it can almost seem that the origin of life is the destiny of these chemicals.

On the other hand, over the course of natural history, the catastrophic extinctions, some of which have almost decimated life on Earth, do seem to be chance on a grand scale. If an asteroid crashed into Earth, so disrupting the atmosphere that the dinosaurs perished in result (Eldredge, Chapter 4), that would be, in the sense analyzed above, an accident, the intersection of unrelated causal lines. On the old supernaturalistic design account, it would seem very ungodly of God to send in such an asteroid. Or even if, as Eldredge prefers, the massive extinctions have terrestrial causes, and many extinctions occur because climatic changes are too rapid for mutations and natural selection to track, we still have the chance intersection of unrelated causal lines, the one at the genetic molecular level, the other that of global climatic systems. It seems unlikely that God would plan that either.

Eldredge, however, is also impressed with how "it has seemed to a number of biologists (particularly we paleobiologists) that Earth's biota is tough, able to rebound in both an evolutionary and ecological sense after even the worst of biotic devastations." (Chapter 3, p. 68). So a full biological and philosophical account is going to have to include this remarkable power of life to persist and develop over the millennia, despite the chances of its destruction. And we will have to include the RNA molecules Cech has discovered that all but inevitably assemble themselves into living organisms. So we are still puzzling over the mixture of chance and order in evolutionary history. I am going to offer an account of a persistent creativity at work within the individual creatures.

When it comes to human life, Ayala tells us that there has been a selection for intelligence, yet, presumably, the increasing cranial capacity in the hominid line, from 400 to 1400 cc, was all due to chance mutations, so chance is mixed with order again. Ayala even thinks that much in cultural life, especially ethics, is a byproduct of this selection for intelligence owing to its survival value. Byproducts are in some sense accidental, and yet all our human scientific capacity for pure science comes out of this cerebral capacity selected merely for its pragmatic survival function. Biological accounts such as these are much more advanced than the old, Newtonian, matter-in-motion model; and we can be glad that these biologists have escaped the reductionist temptation. But I want to ask whether we do not yet need a still more comprehensive explanation.

At this point, we can do well to recall again, with Gilkey, Alfred North Whitehead's warning that possibilities cannot float in from nowhere (Whitehead, [1929]1978, p. 46; Gilkey, Chapter 7). Gilkey worried over how biologists, as well as philosophers and theologians, could arrive at a strong ethical sense if they had only Darwinian biology with which to support such an ethics. The solution, he suggested, was that if nature edges into mind, mind also edges into nature. I want to give a fuller account of that, one that envisions where these possibilities come from, one that reconciles mind with nature, order with chance. Old, supernaturalistic theism does not work, as Darwin and modern biologists have discovered. But we do need an adequate explanatory account. As a biologist, I want to sketch a picture that I believe is philosophically circumspect.

3. *A Postmodern Theism Consistent with Darwinian Evolution*

For science and theology alike, a great and positive contribution that Darwinism makes to our thinking about nature is the role of chance. It closes the door on absolute determinism and opens the door to freedom and choice. Indeed, there is even a sense in which the role of chance makes both life and ethics possible. Chance makes evolutionary natural history possible. We have to understand what chance makes possible and why it makes these events possible. But many, including Darwin, have never gotten beyond the closed door of determinism, be it a supernaturalistic or materialistic.

Hartshorne hit the nail on the head when he said, "There must be something positive limiting chance and something more than mere matter in matter, or Darwinism fails to explain life" (Hartshorne, 1962, p. 210). The recognition of real chance in nature is the first step away from complete determinism. It is a necessary step if we are to recognize any freedom in any creature. In a completely deterministic world, which Darwin seemed to favor, chance becomes purely a matter of ignorance. An event is called chance if the causes are hidden from us. Often these hidden causes involve the intersection of unrelated causal lines. But the meaning of chance that is so important to recognize in evolution and in life is that all is not completely determined. I may walk down a street. Someone else drives a car down that street. I do not look and the person runs me down. That is chance because neither of us planned it. In a world where this sort of chance occurs, there can be degrees of freedom in the entities that make up this world, from protons to people.

Yet if the world consists of a multitude of independent creators, each with its own degree of freedom, what is to prevent the whole from becoming completely disordered? Too many cooks spoil the broth. There must be something positive limiting chance. Darwinism rules out the notion of an all-determining orderer. That opens the door to another concept of ordering. There are only two ways of ordering. One is dictatorial, tyrannical. The other is persuasive. How could a universal persuasiveness be an ordering influence? Only if there is "something more than mere matter in matter" (Hartshorne). What could make matter more than the mere stuff of the mechanists? It is the notion that the individual entities of creation have a degree of freedom to respond or not to respond to the possibilities of their existence and of their future. Such individual entities are, in this sense, sentient. They have a degree of self-determination.

An analogy that can help us here is that of a conductor of an orchestra seeking to orchestrate a bunch of musicians, each one free to play anything at all. The freedom of the players is constrained by the pervasive influence of the conductor in relation to the sheet of music in front of each of them. The conductor is not a dictator. His job is to bring out the best possible response from each player. Freedom of the creatures, on the one hand, and an overall persuasive influence under which they operate, on the other hand, are two components of a postmodern theism. Chance permits this freedom of the creatures, and there is also an order that orchestrates this freedom. Out of the mixing of these two life arises, and evolutionary natural history continues over the millennia, and human life arises, with its diversity of cultures, with its rationality and morality.

In this more comprehensive account, God is the name of the persuasive influence. The term *postmodern* contrasts with the term *modern*, where *modern* refers to the dominant materialistic or mechanistic world view derived from science. This mechanistic world view sees the world as a world of objects and not subjects. It excludes from its consideration the subjective and essentially all that cannot be weighed and measured. It has been highly successful within the limits of the domain where it can be applied. But it leaves out more than it explains, and so we need pass from such an incomplete "modern" view to a more adequate "postmodern" view.

Richard Levins and Richard Lewontin see this clearly:

> The great success of Cartesian method and the Cartesian view of nature is in part a result of a historical path of least resistance. Those problems that yield to the attack are pursued most vigorously, precisely because the method works there. Other problems and other phenomena are left behind, walled off from understanding by the commitment to Cartesianism. The harder problems are not tackled, if for no other reason than that brilliant scientific careers are not built on persistent failure. So the problem of understanding embryonic and psychic development and the structure and function of the central nervous system remain in much the same unsatisfactory state they were in fifty years ago, while molecular biologists go from triumph to triumph in describing and manipulating genes. (Levins and Lewontin, 1985, pp. 2–3)

A postmodern view must tackle those harder issues, and here biology must notice its limits and become philosophically more sophisticated, if it is to escape from the bondage of reductionism. Foremost among the issues that mechanistic science has been unable to deal with satisfactorily is the historical evolution of rational and moral consciousness, as well as the continuing, present relationship between mind and matter. Ayala, Sober, and Gilkey are all struggling with this problem, which lies at the heart of the sociobiological controversy. Monod, Gould, and Dawkins may reject theism, but they have certainly not solved this problem by so doing, nor have they, in their emphasis on the role of chance in natural history, given us an alternative solution. I will attempt a little later to show how this evolution of mind from matter is no longer the intractable problem for a postmodern theism that it has been for a mechanistic science.

But first I must outline the positive picture I propose. There are two central aspects of postmodern theism: (1) the recognition of creativity of self-determination within the individual entities from protons to the first primitive RNA organisms and to people and (2) the universal persuasive influence in the world, which is called God, to which the individual entities respond. These I now consider in more detail.

4. *Creativity Within the Individual Entities of Creation*

In the dominant interpretation of Darwinian evolution, the individual living organisms are treated like machines, devoid of self-motion in any sense and subject only to external forces. Although it may not, at first, seem that organisms are treated like billiard balls, in fact both mutation and selection are in this category. An organism is shaped by what happens to it from without, either by the causal factors in its environment at the ecosystem level or by the causal factors that affect its molecular mutations. So it is not just in physics that matter is passive. Even for the Darwinians, matter is passive, living matter included, being completely determined by external forces acting upon it. Despite the language of chance, these are really unrelated deterministic causal lines, and so there is no principle of spontaneity inherent in living matter.

Whitehead pointed out long ago:

> A thoroughgoing evolutionary philosophy is inconsistent with materialism. The aboriginal stuff, or material, from which a materialistic philosophy starts is incapable of evolution. The material is in itself the ultimate substance. Evolution, on the materialistic theory, is reduced to the role of being another word for the description of the changes of the external relations between portions of matter. There is nothing to evolve, because one set of external relations is as good as any other set of external relations. There can merely be change purposeless and unprogressive. . . . The doctrine thus cries aloud for a conception of organism as fundamental to nature. (Whitehead, 1926, p. 107)

Whitehead is saying that machines cannot evolve. They can only have rearranged parts. Whitehead enunciated more clearly than anyone how the creative evolution of living organisms could not be understood if the elements composing them were conceived as individual entities that maintained exactly their identity throughout all the changes and interactions. He sought to identify both permanence and change in the entities.

Machines change, but they are not self-creative. Machines may remain the same through changes, but they do not maintain any self-identity. Mechanistic identity is not biological identity.

Evolution is not simply the evolution of objects. It is the evolution of subjects. A critical thing that happens over evolutionary history is the change in internal relations of subjects. As contrasted with external reactions, internal relations are constituent of the character and even the existence of something. A similar, if not the same point, is made by Richard Lewontin, Steven Rose, and Leon J. Kamin when they discuss the different levels on which atoms are assembled to make molecules, molecules to make cells, and so on. "As one moves up a level the properties of each larger whole are given not merely by the units of which it is composed but of the organizing relations between them. . . . These organizing relations mean that the properties of matter relevant at one level are just inapplicable at other levels" (Lewontin, Rose, and Kamin, 1984, p. 278). The traditional mechanistic notion of the constitution of the world out of separately existing parts is turned upside down. The whole organizes and even creates the parts. The lower levels of organization are to be interpreted in terms of the higher. It is not just that wholes are more than the sum of their parts. The parts become qualitatively new by being parts of the whole. A cell in the brain is different from a cell in a culture of brain cells.

The point I am making is that cells, organs, and organisms are not at all like nuts and bolts in a machine. They have an inner responsiveness to their environment analogous to our inner responsiveness to our total environment, which includes purposes we choose. That responsiveness comes from the organism as a whole, and it is in this responsiveness that creativity is at work. Another way of saying this is to say that organisms have a degree of self-determination. So, in that sense, they have mind and sentience analogous to what we call feelings in ourselves.

How far down the scale of nature can we suppose that this is the case? Since going downscale is, generally, going backward in time, when did such responsiveness appear? The conventional wisdom is that at some point in the evolutionary sequence from atoms to human beings, mind and feeling appeared for the first time. Something that before was an object without any aspect of mind afterward became a subject with mind. Another way of putting this is to say that mind "emerged." But that simply restates the problem and solves nothing. It only gives a name to something for which we have no explanation. As Sewall Wright said, "The emergence of even the simplest mind from no mind at all seems to me at least utterly incomprehensible" (Wright, 1953, p. 14). I agree with Wright and Whitehead and other process thinkers that no line is to be drawn anywhere down the line from protons to people. This coming together of our understanding about the world around us and the world within us is one of the great insights shared by the new physics and process thought.

In biology, my understanding of molecular genetics is that the billiard ball concept of atoms and molecules is no longer relevant there. Reductionism is outmoded. Genes are no longer regarded as pellets of matter that remain in all respects identical whatever environment they are in. Rather, they are bits of information being creatively used by the organism in an environmental response that produces continual creativity. I think Cech's discoveries about RNA's self-creative activity at the origin of life quite support this view. Further, after life originates, the pathways of gene action no longer have to be regarded as 100 percent determined in any particular situation. They may be 99.99 percent determined, but that difference is all the difference in the world (Birch, 1990, pp. 25–26). This makes creativity possible, because it mixes order with enough freedom to make the world of natural history possible. This is the difference that makes the world.

It is important to distinguish between individual entities that are subjects with their own self-determination and assemblages of subjects such as a rock that have no self-determination. These are properly called aggregates. Mechanistic and reductionist science has been spectacularly successful in this area. The modern world view was based on the study of aggregates such as stellar masses in orbits and steel balls on inclined planes, both of which have no self-determination. Absolute prediction and reduction is possible in principle there. But the mechanistic method has been less successful with the so-called ultimate particles of the universe, such as the electrons, protons, and quarks, which are subject both to indeterminacy and to creative organization into higher levels. It has been progressively less successful with living cells, with higher forms of life such as rats, and least of all with humans (Griffin, 1990, p. 25). The laws applying to aggregates, which are the laws of mechanics, do not provide the ideal for a scientific study of individual entities that are subjects. A scientific study of a world of subjects needs to encompass the subjective as well as the objective aspects of individuals. The new physics, so-called, gets into this area as do some studies of animal behavior such as those by Jane Goodall (1986) and Donald Griffin (1984).

5. *The Divine Eros, Luring and Saving Values in the World*

In the midst of this creativity and self-determination within the individual creatures, there is at work the persuasive influence of God, the divine *eros* (from a Greek word for "love"). The principle of self-determination in the individual entities of creation can be regarded as an appetition, or lure. We experience this principle in ourselves as imagination and the anticipation of as yet unrealized possibilities. This principle is seen as extending to the whole of creation from protons to primitive cells to people. But now we have to ask, to what do the individual entities respond? The first level of answer is to the possibilities of their being, including their future. At the heart of the universe, even before there were cells or atoms, there was the possibility of these entities coming into existence. But at a second level of answer, the potentiality of the universe is conceived as cosmic mind. God is the ground of the possibilities that, in the limited vision of a strict science, has to recognize these possibilities but has no theory for their origin; they can only seem to be possibilities that float in from nowhere.

A religious vision can see further, detecting the origin of these possibilities. Such a reality is recognized, though not explicitly taken to be God, in Buddhism. In Hinduism it is thought of as Brahman. In Judaism and Christianity it is called God. The possibilities of the universe are realities that constitute a continuous lure to creation. This is Whitehead's doctrine of the divine eros or primordial nature of God. The divine eros confronts what is actual in the world with what is possible for it. It is the immensely sensitive outgoing lure brooding over nature. It is the ordering principle at the heart of the universe, else there would be only chaos. It is the ordering principle that uses chance to construct life and human life, and keeps the system viable over the epochs of evolutionary and cultural history. How else can we account for the order of a universe made up of a multiplicity of creative individual entities?

At one level, natural selection is an ordering principle. But it does not account for the order on which it operates, since there is already much order in the universe long before natural selection starts to operate. Nor can we suppose that we simply have that prebiological order by chance. Given that prior order, natural selection itself has to originate at the time that life originates; and even after natural selection originates, there is nothing in natural selection to insure the presence of increasingly complex order. Simple forms of life are naturally selected just as much as complex ones are. A scientific materialism refuses in principle to take this creation of order as a problem. It has to ascribe it to chance. But only chaos would result in a universe of externally related parts.

Only chaos would result, again, in a universe of organisms each pursuing its own ends, with no integration into a community of the whole. Apart from God, there is no way to understand how there could be any limits to anarchy implied in a multiplicity of creative agents, none of which are universally influential. If we may put it this way against the biologists who suppose that natural history is only struggle and the sociobiologists who suppose there are only "selfish organisms," it is God who takes each of these entities that is pursuing its own self-actualizing and organizes ("orchestrates") the impressive natural history that starts with protons and rises to produce first living, replicating cells, and then, eventually, people.

The concept of the divine eros does not mean a supernatural God who could exist apart from any world, who created it out of nothing, who interrupts the natural order of the world, an omnipotent being who completely determines everything. We began this essay by finding that, despite the mechanical world of the Newtonian world view, the world is not ultimately mechanical. Now I conclude by claiming that God is not a divine mechanic. We can be grateful to Darwinian biology for teaching us that the world is not mechanical; nor can God be a mechanic of such a world, for such a God is really external to the universe He has made. God, as Whitehead insisted, is the ground of order and of novelty interacting with our world (Whitehead, [1929]1978, pp. 40, 247).

The divine eros is not before all creation but with all creation, involved as the lure in every creative event. This is not a doctrine of pantheism but panentheism or panexperientialism. God is involved in the world, but not identified with the world. Order is in principle the rule of one. A concept such as the divine eros is needed to account for the remarkable order of the universe of self-determining entities. The divine orderer does not manipulate the world. "What happens," says Hartshorne, "is in no case the product of his creative acts alone. Countless choices, including the universally influential choices, intersect to make a world, and how concretely they intersect is not chosen by anyone, nor could it be" (Hartshorne, 1967, p. 59).

The ordered universe contains within it much that is disordered and incomplete. Multiple creativity makes some disorder and conflict inevitable. It allows for the possibility of disorder and evil. Evil springs from chance and the freedom that chance allows—not from providence. God does not unilaterally determine any events and could not do so. The individual entity's capacity for self-determination cannot be canceled or overridden. The question, then, of why God does not unilaterally prevent particular evils does not arise. The God of postmodern theism does not manipulate things or people.

The God that does not determine the future does provide, at the same time, the possibilities of the future which may or may not become concretely real in the fullness of time. In the past history of Earth, such possibilities have actualized with the formation of living things, with the formation of human beings; and, even today, the future is open ended. There is an overall heading provided by the lure of God, but much is left to local spontaneity. In this way, the creativity of God is increased because God is the origin not simply of His own creativity, but of myriads of other creative centers, to each of which is given genuine creativity. The world view of the divine conceived as a persuasive agency, rather than as a manipulative one, should be looked upon, says Whitehead, as one of the greatest intellectual discoveries in the history of religion (Whitehead, 1942, p. 196). It was plainly enunciated by Plato in his view that ideas are effective in the world and that forms of order evolve.

There is yet a further character to this divine lure, one that is even more speculative and just as necessary. The first fact is that of self-determination within the creatures. These possibilities, their actualization, and orchestration are not fully explained until we detect God as divine lure, in, with, and under these events. Now, further, once value has been achieved, God responds to the creative activities of the world to the extent of saving, as a sort of memory, all of value that has been achieved. From this perspective, this saving of values achieved is the ultimate reason why "Earth's biota is tough, able to rebound in both an evolutionary and ecological sense after even the worst of biotic devastations" (Eldredge, Chapter 3). For the scientist, this is a fact of natural history, but for the theologian, God feels and saves the world. This is the divine passion, or what Whitehead calls the consequent nature of God, which feels the feelings of all creation. This divine love of life is the deepest explanation for the prolific Earth over the last several billion years. This is the literal meaning of panentheism—all things in God.

6. *The Criterion of Harmony*

The question is normally asked in a scientific community, "How testable is this hypothesis?" Testability may well apply to the special sciences when they are dealing with a narrow range of observations and when controls are possible. But such a criterion of testability does not apply to the hugely broad areas of metaphysics. Testability is difficult enough to apply even to theories about the origin of life; as Cech says, we can only ask whether the theories are plausible. Testability does not apply to a general interpretation of natural history, which must involve not only science but philosophy as well.

Many people have concluded that the enterprise of metaphysics is meaningless. They equate meaning with falsifiability (Griffin, 1990, p.101). They like to quote Karl Popper to this end. But I once asked Popper directly if he believes there is meaning to be found beyond the boundaries of what can be tested this way. His answer was, "Yes, of course there is." A metaphysical view can be expected broadly to harmonize with our descriptions of the world and our experiences in the world, but it cannot be put to a direct and critical test. The world, interpreted as a whole, is much too complex to be put to a simple test, although hypotheses about certain phenomena in the world can be tested empirically.

There are other criteria for accepting metaphysical hypotheses. These are tested by their self-consistency and their adequacy to illumine all the known facts. Broadly speaking, this is the principle of harmony. Almost the first sentence in the first chapter of Whitehead's *Process and Reality* deals with this issue: "Speculative Philosophy is the endeavour to frame a coherent, logical, necessary system of general ideas in terms of which every element of our experience can be interpreted. . . . The philosophical scheme should be coherent, logical, and, in respect to its interpretation, applicable and adequate." He then went on to explain his meaning of *coherent*, *logical*, and *applicable*. His analogy helps: "The true method of discovery is like the flight of an aeroplane. It starts from the ground of particular observation; it makes a flight in thin air of imaginative generalization; and it again lands for renewed observation rendered acute by rational interpretation" (Whitehead, [1929]1978, pp. 3, 5).

From aloft, seeing Earth whole as if from space, viewing natural history over the millennia on global scales, some of our observations belong to biological science. Some theories about events going on on Earth can be put to test and be more or less corroborated. But when we move comprehensively across all the possibilities actualized during this startling world history—beginning with the origin of life, continuing with the elaboration and diversification of life, perpetually redeemed in the midst of its perishing, achieving the subjective genius of human moral and rational life—we are lifted into the range of philosophical and theological visions. From where I see it, not only is a postmodern theism harmonious with Darwinian biology, but Darwin even helps to point us in this direction.

Much of the attention in this volume has been turned to the past, to the origins of life, and of human rational and moral life. But we have been turning to the past to find new directions forward, to find what we ought to do and where we ought next to go with this marvelous mind with which we have been endowed. We are trying to understand who we are, where we are, and where we are going. Much is still not clear. But at least this should be obvious. The metaphysician has an ongoing research program that may never end. The possibilities remain open and vast. This came home to me at the end of a conference in Japan on process thought when Charles Hartshorne made the remark that we had before us a research program for a thousand years!

References

Birch, Charles. 1990. *On Purpose*. Kensington, New South Wales: New South Wales University Press.

Born, Max. 1971. *The Born-Einstein Letters*. New York: Walker and Company.

Darwin, Francis, ed. 1888. *The Life and Letters of Charles Darwin* (2 vols). London: John Murray.

de Beer, Gavin. 1973–74. "Evolution." In *The New Encyclopedia Britannica* (15th ed., vol. 7, pp. 7–23). London: Encyclopedia Britannica.

Goodall, Jane van Lanwick. 1986. *The Chimpanzees of Gobe: Patterns of Behavior*. Cambridge, MA: Harvard University Press.

Griffin, David Ray. 1990. "The restless universe: A postmodern version." In K. J. Carlson, ed., *The Restless Earth: Nobel Conference XXIV* (pp. 59–111). San Francisco: Harper and Row.

Griffin, Donald R. 1984. *Animal Thinking*. Cambridge, MA: Harvard University Press.

Hartshorne, Charles. 1962. *The Logic of Perfection*. LaSalle, IL: Open Court.

Hartshorne, Charles. 1967. *A Natural Theology for Our Time*. LaSalle, IL: Open Court.

Hartshorne, Charles. 1984. *Omnipotence and Other Theological Mistakes*. Albany, NY: New York State University Press.

Kingsley, Charles. [1863]1930. *The Water Babies*. London: Hodder and Stoughton.

Levins, Richard, and Richard C. Lewontin. 1985. *The Dialectical Biologist*. Cambridge, MA: Harvard University Press.

Lewontin, Richard C., Steven Rose, and Leon J. Kamin. 1984. *Not in Our Genes: Biology, Ideology and Human Nature*. New York: Pantheon Books.

Passmore, John A. 1959. "Darwin and the climate of opinion." *Australian Journal of Science* 22:8–15.

Tennyson, Alfred. 1850. *In Memoriam*, Part LVI, Stanza 4.

Whitehead, Alfred North. 1926. *Science and the Modern World*. New York: Macmillan.

Whitehead, Alfred North. 1942. *Adventures of Ideas*. Harmondsworth, Middlesex: Penguin Books.

Whitehead, Alfred North. [1929]1978. *Process and Reality: Corrected Edition*. New York: Free Press.

Wright, Sewall. 1953. "Gene and organism." *American Naturalist* 87:5–18.

Endnote

1. I largely depend on Griffin's (1990) interpretation of world views.

Epilogue

HOLMES ROLSTON, III

We have come to an end, and the question is whether there is more at the end than at the beginning—any advance in understanding as a result of our inquiries. In a way, the conclusion our authors have repeatedly been reaching is just such an advance. We have achieved a great deal, if we recognize that, in the natural and cultural history of the world, there is more at the end than there was at the beginning. The authors diversely and collectively give witness to that.

But how do we understand such development? As possibilities floating in from nowhere? First there is A, then B. First the simple elements, then the stars, then complex elements, then a prolific planet. First energy, then matter, then life. First simple life, later complex life. First singular identities, then fused symbiosis and collective identity. First primates, then persons. First genes, then ethics. In the beginning, physical matter; in the end, metaphysics. Is this B unfolding from A? A as precursor, necessary but not sufficient for B? More out of less?

Life and mind are known only on Earth, where life has been abundant across much of geological history, with a turnover of several billion species across several billion years. Mind—in the sense employed here: the human self-conscious, reflective mind—is rare even on Earth. Hardly any of the other billions of species even came close, remembering that most were or are plants, protozoans, crustaceans, insects, and the like. Among the mammals, in about twenty orders, only the primates developed possibilities of self-conscious minds, and among a hundred species of primates, only one, *Homo sapiens*, reached the level of mind that can do science and ethics. In the preceding pages, some of the keenest of such minds have been reflecting over our origins, the origin of life, of mind, of ethics.

1. *Thomas Cech: Catalysis and Creativity*

It is difficult to avoid the conclusion that the natural history on earth is both natural and startling. Thomas Cech draws a rough analogy between the organism with its DNA coding for functional metabolisms and a videotape recorder, on which coded tapes can be played. Later, he portrays one kind of molecule, RNA, that is both tape and player and self-assembling. Analogies always have their limitations, but if we try to push the analogy here, we should think it rather remarkable if there were a videotape that folded itself into a VCR player and reproduced itself, and we should think it even more remarkable if we learned that this video-tape-cum-VCR-player was self-assembling out of bits and pieces lying around the laboratory. More remarkable still if it developed a mind and a conscience. Perhaps that is just evidence that the VCR/videotape analogy breaks down; but it may also be evidence of how dramatic a form of creativity did operate in the set-up of those primordial seas. We have no reason to think, Cech says, "that therefore life ceases to be an extraordinary thing, and becomes an ordinary thing" (p. 32).

Catalysis is the biological word for this; catalysts make otherwise improbable things happen. But in a more philosophical perspective, this yields an amplified synthesis of forms, increased possibilities, generation, and creation. "*Catalysis* . . . is the right word from a technical biological perspective; but what is also involved here is what a philosopher or a theologian might call creativity" (pp. 32–33). What catalysis does, to keep the big picture on the horizon, is to bring new possibilities and make them actual. So the subsequent story of life becomes all the more complex.

Another conclusion here is that getting more out of less requires a mixture of the inevitable and the contingent. "We see life as being the likely, perhaps even the inevitable consequence of chemistry" (p. 31). Cech finds the possibilities of life intrinsic to the chemical molecules. There is *autocatalysis*, self-organizing. The chemicals of life are common ones—carbon, oxygen, hydrogen, nitrogen—not only on Earth but elsewhere in the universe, although the geological conditions on Earth are unusual. So there is something required not only by way of the chemicals, but of their special opportunities on Earth, as Sagan and Margulis insist.

Cech's account certainly makes the possibilities more probable, although it does not reduce the need for a philosophical account of the origin of those possibilities. The Earth materials whose chemistries Cech, as a biologist, takes as a given had to come from somewhere. But that is a question for astrophysicists and geologists, for cosmologists and metaphysicians, more than for biochemists. A fuller explanation will need to incorporate what such disciplines are now saying about the anthropic principle and the "fine-tuned universe" (Leslie, 1989).

Cech is not the first Nobel laureate to find life intrinsic to the chemicals. He draws attention to the work of Melvin Calvin. Life "arose not by accident but because of the peculiar chemistries of the various bases and amino acids. . . . There is a kind of selectivity intrinsic in the structures" (p. 31). We can add others. George Wald asserts: "This universe breeds life inevitably" (Wald, 1974, p. 9). Manfred Eigen concludes "that the evolution of life . . . must be considered an *inevitable* process despite its indeterminate course" (Eigen, 1971, p. 519).

On the other hand, there are many, Nobel laureates among them, who think that life is an accident. Jacques Monod finds evolutionary natural history to be "pure chance, absolutely free but blind . . . an enormous lottery presided over by natural selection" (Monod 1972, pp. 112, 138). And, from the point of view of pure physics and chemistry, there really is not much, if anything, in the electrons and protons, in the atoms *per se*, that suggests they will organize themselves into microbes, much less insects and dinosaurs. The natural history from protons to protozoans to people, as Charles Birch puts it, needs some "lure." Those possibilities are there, but still, when they come, they come as quite a surprise. If we were to run the story over, we would not get the same story. So we hardly know what to say. On the one hand, we are impressed with the possibilities intrinsic to the chemicals; on the other, we are impressed with the contingent narrative of events in which—despite death, extinctions, and catastrophes—there are ever novel achievements of biological vitality and power.

2. *Dorion Sagan and Lynn Margulis: Facing Earth's Symbiosis*

Before we can get more out of less, we must have a suitable planet. The planet must be large enough to retain an atmosphere; it must have cooled down to a temperature that accommodates life. It must be within a congenial range of the energy poured out from the star it orbits, near enough for the supply of warmth to be adequate but not so close as to make the heat intolerable. The orbit needs to be approximately circular or the temperatures between perihelion and aphelion will be extreme. The orbit must be stable over long epochs, not disrupted by collisions with or near approaches to other planets or stars. The planet's rate of revolution must be such that there is no excessive heating in one hemisphere and excessive cooling in the other.

Although many liquids are possible on planets, water is the really exceptional substance with anomalous properties, on which many of life's properties depend, such as its solvent capacities. At least the only form of life we know depends so. So there must be fluid water, lots of it. The fluid water must be driven to circulate. If there is to be terrestrial life, there must be land masses, adequately irrigated by the atmospheric circulations, and these land masses need to be periodically rejuvenated from erosion. There needs to be a perpetual churning of the materials, irradiated with energy. The atmosphere, waters, and lands must have the important precursors for the synthesis of organic materials. On Earth, what would otherwise be extremes of heat and cold, wet and dry, are moderated by the circulations of water.

All these circumstances are present on Earth, we might say, by cosmic accident. That does not mean that there were no causes that resulted in these conditions; to the contrary, there were causes for all of them. But the result is an unplanned concatenation of many otherwise unrelated causal chains. Earth is a very lucky planet, a quite special place. George Wald's universe that breeds life inevitably is too grandiose a claim; the universe only rarely breeds life. The most we might say is that "Earth breeds life inevitably." So it does take luck after all to get much history; most of the planets of our universe, like the other eight in our solar system, like the ten thousand asteroids, will be stillborn. A planet right for life is one in a million. But after that kind of luck, as Cech has discovered, life is highly probable, even inevitable—not so much intrinsic to the chemicals as intrinsic to the set-up.

Earth is a remarkably different planet not just by astronomical accident but also by biological take-over, so that today, if one compares Earth with its sister planets Venus and Mars, not only was life once possible here, while impossible in the heat of Venus or the chill of Mars, but life has become prolific here because Earth has been rebuilt systemically. The mix of gases here is not what it is on Venus and Mars, and not what it once was on Earth. Terrestrial temperatures have stayed relatively constant, while the sun's luminosity has increased by a quarter. Ocean salinity has remained at less than 10 percent of saturation for millennia. For Sagan and Margulis, such global features mean that the phenomenon of life has to be seen from the top down, not just from the bottom up.

The more out of less comes from synthesis and symbiosis. Life begins as microbes, but a repeated symbiotic fusion of identities has characterized life on Earth, from the microbial on up to the planetary levels. One identity can flow into another; identities can be joined to produce a new identity at a higher level. On Earth there are "formed new kinds of more complex 'selves' with unique identities. One 'self' comes to merge with another 'self' and a new 'self' arises. Biological identity is not fixed. . . . Individuals join to produce new identities at more inclusive levels. This kind of merging to form larger individuals is the way of the world" (pp. 59–60). Even in human society we take our identities significantly in terms of the roles we play in the traditions we serve. The highest of such synthetic or collective levels that we face is Earth itself. Gaia is the explanation for the development of life on Earth, at least for its natural history. In their cultural history, in their moral history, humans may have to forget and to transcend their amoral roots in nature, but they cannot and ought not to forget their Earth. We are facing a system with a programmatic fecundity. That brings responsibility for the biosphere, for the community of life on Earth.

3. *Niles Eldredge: Life Hanging Tough*

Life develops over the millennia, but with many vicissitudes, misfortunes as well as fortunes. Getting more out of less is connected with these ups and downs. Life moves up, but often there is a downside before an upside; the upside is life rebounding after tragedy. Upset and rejuvenation are what make speciation possible, really novel speciation by which life can advance. The pattern is that the big changes, including the advances, come after the environmental stresses that result in extinction.

Eldredge concludes: "The particularly compelling aspect of this account is that the factors underlying species extinction—namely, habitat disruption, fragmentation, and loss—are the very same as those conventionally cited as causes of speciation. Thus the causes of extinction may also serve as the very wellspring of the evolution of new species" (p. 79). There is, for instance, a step up of mutation rates under stress. Species evolve most rapidly under conditions where environments change most severely. Extinction is omnipresent, sometimes catastrophic, but, in the end, extinction is not the rule after all, because here we humans are, with millions of species in our care. Paradoxically, the end of extinction is humans with an ethic for the conservation of life.

David M. Raup, the paleontologist who, with Eldredge, has most extensively explored extinctions, also holds that these periodic cutbacks prepare the way for more complex diversity later on. Were it not for extinctions, including catastrophic ones, we humans would not be here, nor would any of the mammalian complexity. Life on Earth is so resilient that normal geological processes lack the power to cause widespread extinctions in major groups. But just such a resetting is needed—rarely but periodically. We should think twice before judging catastrophic extinctions to be a bad thing.

Raup explains:

> Without species extinction, biodiversity would increase until some satura-
> tion level was reached, after which speciation would be forced to stop. At
> saturation, natural selection would continue to operate and improved
> adaptations would continue to develop. But many of the innovations in
> evolution, such as new body plans or modes of life, would probably not
> appear. The result would be a slowing down of evolution and an approach
> to some sort of steady state condition. According to this view, the principal
> role of extinction in evolution is to eliminate species and thereby reduce
> biodiversity so that space—ecological and geographic—is available for in-
> novation. (1991, p. 187)

There is a big shakeup that is in some sense random; it is, we must
say, catastrophic, but the randomness is integrated into the creative system. The loss of diversity results in a gain in complexity. Catastrophic extinction "has been the essential ingredient in the history of life that we see
in the fossil record" (Raup, 1991, p. 189). The storied character of natural
history is increased. Once "we thought that stable planetary environments
would be best for evolution of advanced life," but now we think instead
that "planets with enough environmental disturbance to cause extinction
and thereby promote speciation" are required for such evolution (Raup,
1991, p. 188). So the extinctions are required to produce more out of less.

This extinction/evolution coupling is part of our own origins. Eldredge finds this in human evolution too: "Thus the pattern holds: *Homo
erectus* appears to have evolved in conjunction with a major event of climate change" (p. 79). That is philosophically as well as biologically interesting: the churn that destroys life creates life. We may not yet be sure
whether possibilities are floating in from nowhere. But it looks as if when
they come, they come in crisis events.

From this point on, we have increasingly to reckon with the human
phenomenon, where there is, strikingly again, the evolution of more out of
less. During our human evolution, in an epoch of climate crisis, culture
evolves, and, rather more recently, agriculture evolves. "Advances in culture
undoubtedly underlay the increasing ecological flexibility of *Homo erectus*
over its forebears" (p. 80). These new possibilities continue in the ever-
accelerating capacity of *Homo sapiens* to remake habitats, owing to the transmission and accumulation of cultural skills, agricultural skills over several
millennia, and, over recent centuries, industrial skills. "It is through the activities of *Homo sapiens* . . . that the world's biota faces its current crisis. . . .

We are very much responsible for the present-day biodiversity crisis" (p. 67). True, there may be global climate change, global warming, but this time we with our agriculture and industrial behaviors—and not astronomical collision or spontaneous natural climate change—are the causative agents.

With these new powers comes moral responsibility, something more once again. "We as individuals were born into, and as a collective species *evolved* in, a natural world that we ought to feel compelled to restore and protect" (pp. 68–69). But how do we get from that fact of our evolution to that interspecific duty to protect and restore it? How do we even get, much less authorize, duties to care for our fellow humans? Eldredge is cautious because he knows biology too well, and he knows how theories in biology have shifted over time, having himself been part of such shifting paradigms. He is dissatisfied with philosophical efforts to derive what *ought* to be in ethics from what *is* in biology. So he is reluctant to derive more than self-interest out of biology and hesitant to say where his ethics, over and above that, come from. "Thus I return to human responsibility: responsibility, if to no others, at least to ourselves, our own species" (p. 83). Fortunately, at the global scale, humans have entwined destinies with the natural world with which they coinhabit Earth.

So Eldredge, too, insists that there are genuine novelties in the evolutionary story, periodic upsets and renovations all along, and, especially when humans come, startling novelties. Culture is one, and agricultural culture is another; technological, industrial culture is yet one more. Ethical agency is still another. Perhaps a cautious biologist is well advised to be leery of deriving any of these things—culture, agriculture, technology, industry, or ethics—out of mere biology. This reluctance we can take to be more evidence that there is more out of less, emergent phenomena that biology cannot explain. Nevertheless, a biologist does have to confront all the dimensions of the *humanum*, Gilkey's human world, and we will have to have some account of the origin of these phenomena. If not, we do have possibilities floating in out of nowhere.

4. *Michael Ruse: Ethics as Authentic Illusion*

From midway through our anthology onward, we have had to face ethics head on. Michael Ruse is excited. "What excites the evolutionist is the fact that we have feelings of moral obligation laid over our brute biological nature, inclining us to be decent for altruistic reasons" (p. 97). But Ruse is reluctant to find ethics to be much more out of less; he wants to unfold it from biology. Biology is displayed unusually in the vicissitudes of human culture, ethics included, but displayed without gainsaying natural selection. This, however, is not a claim to be feared. Fortunately, what natural selection requires in culture is a quite cooperative and agreeable ethic. There is no need to free biology from ethics, much less to set biology against ethics.

Ruse does not want possibilities floating in from nowhere either; and, lacking a source outside Earth's evolutionary history from which to originate those possibilities, Ruse has to get his more from less by unfolding biology. But that forces him to twist and turn, because he does need a place to legitimate ethics, and the obvious candidate is in a rational justification by social contract. We do not have to go outside Earth for the origin of ethics (not to some supernatural God), but perhaps we do need to go at least to culture, a culture on Earth but that transcends nature, notably so when ethics appears.

This struggle is revealed in Ruse's appeal to John Rawls' concept of justice. Justice is what a person would wish in society, if he or she were blind to the particulars of his or her own advantage, and, behind that veil (in what Rawls calls the original position) if he or she acted in terms of self-interest. That would include fair treatment at the hands of others in a society with sufficient moral fabric to make cooperation dependable, and with an equitable distribution of resources. From this arises the social contract, hitherto and otherwise unknown in animal behavior. So ethics both evolves and emerges, more out of less.

But Ruse holds that the self-interest at work in biology is compatible with the reflective self-interest Rawls believes that thoughtful persons would favor, if they were setting up the rules for human cultural life. Ruse recommends the same behavior that Rawls recommends, but Rawls finds more rational justification than can Ruse. "Ultimately, there is no reasoned justification for ethics, in the sense of foundations to which one can appeal in reasoned argument. All one can offer is a causal argument to show why we hold ethical beliefs. But, once such an argument is offered, we can see that this is all that is needed" (p. 101). Rawls has gone to great effort to reason out an argument as to why cooperative morality is the right, as well as the best, course for humans. But Ruse concludes: "The evolutionist's case is that ethics is a collective illusion of the human race, fashioned and maintained by natural selection in order to promote individual reproduction" (p. 101).

But can the biologically self-interested person who has been relocated to Rawls' original position really act in self-interest any more, when one does not know which or what kind of self one will be? The particularities of one's historical self are stripped away. I must choose for the good of a generic self; my later historical interest (my particular genetic set, my particular history) is unknown to me. It is hard to be selfish if there is no particular biological identity to defend, if one is choosing in a context where one could turn out to be anybody, anywhere, anytime. One doesn't know one's coding, nor one's coping. What is left is a choice that does not seem "particularly" selfish any more because it does not know any of the particulars of individual life, though it defends "me" universally, wherever I might be located. This is selfishness interlocked conceptually with the fate of others, forced in that sense to be altruistic (other-minded) from the start.

If this is enlightened self-interest, it is quite widely enlightened self-interest, since it defends one's interest without knowing who one's self is at all, nor how to be particularly interested in it. The self is smeared out by the veil of the original condition. It is defending one's unknown self-interest which could be anybody's interest, since I might be anybody. In "the original position of equality" the self is equally the other, unable to differentiate itself. Perhaps this is more of what Sagan and Margulis identified as the symbiotic fusion of identities that has characterized life on Earth. We are now meeting this symbiosis in moral life, where self-interest cannot go it alone but must synthesize with the interests of all others. That can just as plausibly be interpreted as humans reaching a more that cannot be reduced to the less of biology.

"Free, equal, democratic." "Fair, just, loving." This morality no longer sounds very selfish, although it does defend human welfare generically. We do, after all, want to defend the interests of "selves." Persons are valuable achievements. This kind of universal defense of human welfare is no bad thing. If we can succeed in justifying rationally the moral inclinations we have inherited genetically, well and good. We might then have to ask, however, whether, limited to evolutionary theory alone, such justification is a possibility that has floated in from nowhere.

Ruse fears a transcendent "more," infusing the natural and telling humans what to do. But, actually, there is no clear reason yet stated why such ethics could not also be divine. The commandments were given, in the Biblical account, in order that the Hebrew people might inherit their promised earth. Perhaps morality is given on Earth in order that we humans can inherit the Earth—if and only if we do so in justice and in love. We "throw help into the general pool and draw freely from it as needed" (cf. p. 104). There is nothing so ungodly about that. To the contrary, it sounds much like what early Christians recommended as an almost too-idealistic ethics of love (Acts 4.32–35). Charles Birch finds this ethical development to be the lure of divine love.

Once upon a time, a population of primates evolved into ethical animals that now read Rawls on the foundations of justice, worrying about altruistic love and the Golden Rule, about the greatest good for the greatest number. This ethics that we humans in fact enjoy seems quite underdetermined by the biology we inherit—as all the remaining authors, Ayala, Sober, Gilkey, and Birch agree. And if our ethics is underdetermined by our biology, then we must cast around for where the possibilities come from—perhaps in something transcending biology.

5. *Francisco Ayala: Necessary and Sufficient Byproduct*

The origin of mind is as much a puzzle as the origin of life. The real puzzle is how to get subjective self-consciousness (people) out of merely objective biology (protozoans); Birch indeed holds this to be impossible; too much from too little. But the origin of mind, others reply, is not so implausible because having a mind is evidently adaptive. Smart animals cope better in the world. Yes, but the first trouble with that reply is that most animals who do well in the world are not smart in the human cognitive sense at all; indeed, the most successful animals (the Coleoptera) do not have minds at all.

The second trouble with that reply is that so much of what humans do with their minds does not have any evident survival value. Intelligence was targeted for such dramatic cerebral increase (400 to 1400 cc), though the increase of cerebral power is really needed for only a small subset of the larger activities of which this big brain is capable. The human mind seems oversized for its evolutionary role. In the case of monkeys, coyotes, and deer, their brain powers are close-coupled with their survival and reproductive behaviors, but in the case of humans, a brain/mind was targeted that can serendipitously do these many other things as well. Natural selection is a cause of human cerebral evolution, although not the explanation for most of the powers of the brain. They come as what, from the point of the view of the biological theory, have to be described as byproducts.

Curiously, natural selection has to build a brain all but a subset of whose creative activities lie outside the scope of natural selection. We can and ought to act with what Sagan and Margulis called an "active forgetting" of our biological past. Such a brain has to enable its owner to reproduce, but, beyond that, there are a host of what Sober calls free-floating possibilities, one of which is morality. For Ayala finds ethics to be unrelated to survival. "I see no evidence that ethical behavior developed because it was adaptive in itself. I find it hard to see how *evaluating* certain actions as either good or evil . . . would promote the reproductive fitness of the evaluators" (p. 122).

Before we try to answer whether ethics is nonadaptive, we should recall Ayala's longer list of nonadaptive byproducts: literature, art, science, technology, politics, and religion, as well as ethics. That is a rather comprehensive list; we first wonder what activity of intelligence is not on the list. Ayala does not mean, of course, that science and technology are never adaptive, for he gives evidence to the contrary. Technology provides axes to cut trees; science provides medicine to cure childhood diseases. He means that some forms of these activities have nothing to do with family life and reproducing. Einstein's theory of relativity did not help him reproduce better. Certain forms of technology—smelting silver for jewelry or making widgets—have no direct bearing on reproduction; humans are always "making ever more complex tools serving remote purposes" (p. 121). Some science and technology is biologically irrelevant. In *Homo sapiens*, we can easily cite activities in the fine arts, or pure sciences, or politics, or religions that do not immediately to bear on reproducing; these are byproducts.

On the other hand, literature, art, politics, and religion often do bear on reproducing, in the sense that where these cultural activities flourish, people flourish. Literature and art regularly deal with love between the sexes. Politics is as practical as technology, and religion is a way of coping in the world. Politics organizes communities for better social interactions and tribal and national defense; religion urges parents to care for children. Now we begin to wonder about ethics; it also seems to belong in the array of cultural activities that often promote survival, if also sometimes not.

Ethics also organizes communities for coping. The commandments, given on Mount Sinai, are "that your days may be long in the land which the Lord your God gives you" (Exodus 20.12). No doubt we can find some political, religious, and ethical behaviors that seem irrelevant to reproductive survival (anointing the king with oil, or not mixing meat and milk products at the same meal), but politics, religion, and ethics are often quite as beneficial as technology in helping people to live well in a prospering society and rearing another generation to reproductive age. Ethics integrates into the cultural package of social, political, technological, literary, and religious activities that give humans an "inclusive fitness," if we may adapt a sociobiological term.

The problem with the origin of mind is not that we do not cope with it, but that we do so much more. The problem with ethics is not that we do not cope with it, but that the criteria we use are not found in biology. On that latter score, Ayala is quite right. The criteria are something more than biology. They come from culture. This forces him to posit ethics as a byproduct. But Ayala, in turn, forces us to ask whether, after this explanation of ethics as a byproduct, explanations are over. Perhaps this is only the end of biological explanation, opening up, by the very appeal to cultural criteria, deeper questions about culture transcending biology.

Intelligence generically is selected for, but the specific activities of intelligence are not. A byproduct implies that something else is the main product. But when we ask what the main target is, we are not given something else that is specific but moved to the level of generality. Intelligence generically, nonspecifically, is the product. All the specific activities of intelligence are byproducts, except where intelligence is specifically promoting survival. Now Ayala is noticing, what seems evident, that intelligence generally can contribute to reproduction, and it is plausible to hold that it was selected for on this account. But intelligence has many uses; some are reproductively clever; many others have nothing particularly to do with reproduction. Natural selection did "target" a specific product of intelligence: getting mates and rearing children. Many other specific intellectual activities of humans, such as writing novels or reading books on ethics are byproducts. Perhaps these activities are good, useful, artistic, moral, practical, theoretical, or not, but they have no adaptive advantage or disadvantage.

Now it is becoming troublesome to say whether ethics is or is not biologically required. It seems to be more than biology. Ayala's starting point is that generic intelligence is the product, with morality entirely a specific byproduct. There is no adaptive value found in judging something right or wrong, although there is much adaptive value in evaluating outcomes of action. But, given Ayala's recognition that the conditions for advanced intelligence are necessary and sufficient for, because logically identical to, the conditions for conscience, he recognizes that morality in general is required by our biological nature, although the content of that morality is to be culturally evaluated. All morality is byproduct of selected-for intelligence. Meanwhile, when we get one we get the other inevitably. After that, though there is inevitably this byproduct, ethics, it is further true that what ethics we shall have is open to cultural evaluation. We worry whether, with such an array of impressive byproducts, we are not overwhelmed by possibilities floating in from nowhere. Only biologically speaking are these all byproducts; in a more comprehensive world view, with a metaphysics, we have to say more.

6. *Elliott Sober: Free-Floating Monkey Wrenches*

From one perspective, Sober offers us a humble, down-to-Earth explanation. "Cultural selection can be more powerful than biological selection. The reason for this is not some mysterious metaphysical principle of mind over matter. When cultural selection is more powerful than biological selection, the reason is humble and down to Earth" (p. 156). And yet that is a surprising discovery. "The astonishing thing about the human brain is that it has brought into being a selection process of its own. The brain is able to liberate us from the control of biological evolution precisely because it has given rise to the opposing process of cultural evolution" (p. 151). "Mind and culture fundamentally alter the format that one should use in explaining behavior" (p. 160, in note 3).

Humans may be descended from monkeys (primate ancestors), but the brain is a "monkey wrench" with "many side effects" (pp. 143, 156–157). "Biological selection produced the brain, but the brain has set into motion a powerful process that can counteract the pressures of biological selection. The mind is more than a proximate mechanism for the behaviors that biological selection has favored. It is the basis of a selection process of its own, defined in terms of its own measures of fitness and heritability. Natural selection has given birth to a selection process that has floated free" (p. 158).

Sober locates the novelty in tempo of change. "Factors that affect evolution are stronger or weaker, according to the amount of change they bring about *per unit of time*" (p. 156). The tempo change that mind introduces, faster by more than an order of magnitude, is the chief explanation how and why cultural cybernetic processes float free. That is certainly important in the explanation, but is there more? Is it only tempo that makes the critical difference in this astonishing power that brain networking has over genetic networking?

Bacteria can reproduce genetically as fast or even faster than humans can successfully transmit ideas culturally; nevertheless, humans, in medically skilled cultures, have learned to control infectious diseases with much success. Speed is not the only criteria: humans can and have deliberately sought cures for bacterial diseases; in contrast, bacteria cannot deliberate about how to evade antibiotic drugs.

Then again, from another perspective, within the peoples of *Homo sapiens*, cultural transmission retains more linkage to the pace of genetic transmission than might first appear. *"Thoughts spread faster than human beings reproduce,"* insists Sober, with emphasis (p. 156). No one wishes to deny this, but if we consider the reproduction of a culture, that still has to be done intergenerationally. One must have babies to indoctrinate them; babies can be conceived in a few minutes, delivered to birth in nine months. But it still takes more or less a generation to educate children to maturity, fitting them culturally to reproduce grandchildren in turn. The transmission of heritable value systems takes the two decades (more or less) of parental dependency; and, in traditional cultures, value systems, which are the real determinants of behavior, do not change all that fast.

Perhaps there is more to consider than just the differential in tempo. Ayala finds that the human intelligence, targeted by natural selection, results in the ability to anticipate the consequences of one's own actions, to make value judgments, and to choose between alternative courses of action. Those are not just tempo differences; they make behavior quite labile because of the novel mode of introduction and evaluation of options. Sober says, "Ideas can plug into a network in which brains are linked to each other by relations of mutual influence. This is a confederation that our brains have effected" (p. 157). But perhaps it is not simply the speed, but the kind, of "relations of mutual influence" that makes the critical difference.

Within cultures, ideas spread in part because they are true. The format changes from survival as the only criterion, in nature, to whether ideas are rational, sensible, plausible, in science, or in ethics to norms asking whether conduct is just, fair, loving. Tempo and mode of transmission can be compounded, but they ought not to be confounded. Genes are transmitted by having babies; ideas jump from head to head. One might first think that the difference in mode is important only because it produces a differential in tempo. But genes are selected, in spontaneous nature, because they have biological survival value; ideas are adopted, in culture, because some heads evaluate the ideas that are jumping around and select those they find to be true, or good, or beautiful, or moral ideas. These may or may not also have survival value.

Ideas spread because they are persuasive in various ways, some of which are rational and logical, some of which are moral and religious, some of which are attractive in other ways. They make life more comfortable, they entertain, they satisfy our curiosity, and so forth—although the only ideas that can stay around over many centuries must allow reproductive success biologically, as well as have a certain appeal psychologically and logically. The real difference is the mode of information transfer; genetic in one case, ideational in the other.

Sober has given us evidence of the more than comes with mind over biology. But he does not go far enough. We need to ask whether there is still more at issue here than genetic versus ideational speed of transmission. The criteria of selection differ: survival of "selfish genes," selection for my "inclusive fitness" drops out or, if present, is overpowered. The criteria of selection are universal human rights. It is not just that thoughts spread faster than genes; thoughts spread in radically different ways than genes, and on different principles. If the end of the story reveals anything about its beginning, this mind edging into matter does affect our conclusions. Byproducts and monkey wrenches are not very helpful analogies if one is seeking a more comprehensive explanatory scheme, complete with a metaphysics.

7. *Langdon Gilkey: Choosing Ourselves: Biology, Biologists, and the Humanum*

If A issues in B, we may, in some circumstances, take A as foundational and explain how B results from A, interpreted in terms of the categories through which we understand A. Given the positions of Earth, the sun, the moon, and gravitational pull on the oceans, we explain how tides result. But in other circumstances, we may take A as a developmental stage of B, in which case A is understood as a preface to B. The acorn, A, produces an oak, B, because that is what it is set to compose. The one is a reductionist account, the other a compositionist. In the evolutionary movement from nature to culture, do we want a developmental or a foundational explanation? "Genetic" explanations are as likely to be developmental as foundational. Acorns (with genes in them) are to be understood in terms of oak trees, as much as oak trees being understood in terms of acorns.

Scientific explanation looks for general theories and initial conditions, then follows a causal chain to find the results. The end is understood in terms of the beginning. But narrative explanation follows story lines; initial events may or may not develop to historical conclusions. Now, more than even with an acorn unfolding into oak, there is possibility becoming actuality with surprises and contingencies, emergents, critical discoveries, novel information appearing, and, in human affairs, resolutions, intentions, and decisions. Where there is development, the beginning has to be understood in terms of the end.

During Earth's history, there has taken place an evolutionary movement from nature to culture, or, equally from matter to spirit. Biology is already itself a historical science, with a story to be told in which more develops out of less. That story does not stop in biology but continues into culture. We certainly wish to know the antecedents of human mentality as these were produced across evolutionary history, and how this "resulted" in culture. But if we put that sequence into a narrative framework, we want to interpret the beginning in terms of the end, not the other way round. We want to interpret the less in terms of its development into the more.

The latter has to be done with care; for if we take emergence seriously we cannot explain the latter in terms unavailable earlier. Thus, I am not going to understand guilt and forgiveness, which appear in humans, by studying trees and flowers, where neither has emerged. Similarly, I am not going to understand trees and flowers by studying rocks and minerals, where photosynthetic life has not yet emerged. There are no genes at the start, unfolding all along, but there is historical genesis. So neither am I going to understand "nature" by remaining forever in physics and chemistry. Or culture by trying to remain in biology.

Explanation from below gets the explanatory sequence backward. It assumes that biology is the higher order premise, from which the lower order phenomena (cultural behavior) can be shown to follow. This is like assuming that physics is the still higher order premise, from which the lower order conclusion (biology) can be shown to follow. But, in fact, the culture follows the biology historically—not inferentially but dramatically. Culture is the higher order, later achieved phenomena, which is created out of, superposed on, and transcends the lower, foundational phenomena, biology. The big picture is historical, not reductionist. I do not know all about atoms and chemistry until I know what they become in biomolecules. Nor do I know all about nature if I remain in biochemistry but must advance to psychology and the capacity for felt experience.

Story explanation is always of this form; in narrative, the later events interpret the former.

In an anthology on biology, ethics, and the origins of life, we might think that what we ultimately need explained is biology and the origins of life. But not so. What is nonnegotiable is the self with its choosing. Explanations in biology, or physics, are not over until they produce a world in which this self is possible. "The call to 'choose ourselves,' to embody in our existence what is given to us, to embody it by inner decision, affirmation, commitment, and perseverance, is one of the characteristics that makes us human; it is the source of responsibility, and so of the moral. . . . This choosing is the referent for the category of 'spirit' and 'freedom.' This requirement to choose ourselves includes the possibilities given us by our genetic inheritance" (p. 169).

What biology insists on, rightly, is that such choosing arises out of evolutionary natural history. "All of the facets of 'spirit' or 'reason,' the entire *humanum*, stretch back into the dimness and mystery of so-called matter, into the mystery of nature as the source and ground of all that we are. . . . If the mind must be understood bodily and naturally, nature as the source of human being must also be understood compatibly with those terms through which we understand human being, including mental, moral, and spiritual aspects" (p. 170). Nature edges into mind, but mind edges into matter.

If this is too theoretical, consider Gilkey's *ad hominem* application to drive his case home. *Ad hominem* arguments are not fallacious but rather demanded when one offers a theory that includes oneself. Is ethics merely biology, or is there something more? One way to test the claim is to make it self-referential. If we try to find, from within biology, where the biologists themselves obtain their higher morality, we find no explanation at all. Worse, we cannot find any grounds for rational evaluation even of the science that is proposed. We have no explanation of how moral conscience and objective reason might be related to the immensely selfish and deceptive morality and reason described in the theory itself. A scientific inquiry is not up to evaluation of this experienced sense of responsibility. But the latter is quite as much both possibility and actuality as is the empirical science of natural history. Before explanations are over, we must include this high sense of morality, as well as any all-too-human lapses into selfishness, a story in which sociobiologists, like all humans, find themselves entwined. Biologists themselves, worrying about ethics, are the best proof of the fact that, when humans arrive on Earth, we have reached more out of less.

8. *Charles Birch: The Lure of God*

First matter and energy, then life, then human life. First protons, then protozoans, then people. By some accounts, all this is unfolding of potential, life intrinsic to the chemistries. The more is all coded into the less. By other accounts, there is no coding at all; there is just luck bringing more out of less. "Almost every interesting event of life's history falls into the realm of contingency," concludes the paleontologist Stephen Jay Gould (1989, p. 290).

Birch suspects that a still better answer is something metaphysical behind the biology, behind the physics, in, with, and under it. It seems equally obvious, over that course of natural history, that the biology does not show sheer contingency; rather, there is, in the contingency, a pattern of life "hanging tough" (in Eldredge's metaphor) over the millennia, becoming more diverse and complex, catastrophic extinctions or not. That is evidence of a prolific principle mixed in with the extinction turnover. If you put it all down to luck, and view three billion years of natural history as nothing but chance riches, then you really do not have any explanation at all. Luck is not an explanation—certainly not of a long-continuing natural history from protons to people. It is a confession that one does not yet have an explanation; it is reason to look beyond the incomplete biological theory to see what metaphysics is most plausible.

On the other hand, we cannot say that all the creative natural history was predetermined. That was the error of Newtonian biology, which often coupled with the error of classical and modern theism. This viewed nature as a deterministic machine, and God as the supernatural Architect-mechanic-carpenter. There is too much truth in the contingency claim to believe this any more about either nature or God. There is chance present in the world, both genuine and relative. Biology teaches us that, and incorporates such chance into the theory of natural selection.

The nonnegotiable component of human world, which must be explained if we are to have the ethical responsibility we undeniably have, is subjective choice. How are we to derive that from mere matter, or even from mere life? Birch does not find that nature is its own explanation, if we suppose that we can start with matter and work this up into egos. One cannot evolve subjects out of objects, any more than one can evolve the number one by summing up and integrating a long chain of zeros. "The emergence of even the simplest mind from no mind at all seems to me at least utterly incomprehensible" (pp. 208–209). Matter has an "inner responsiveness . . . analogous to our inner responsiveness" (p. 208); such a panpsychism is necessary if we are to unify the movement of natural history through the origin of life on through the origin of human life.

On the other hand, though matter has, fundamentally, this psychic component coupled with the physical component, we cannot say that a creative naturalism is an adequate account. There really is not all that much in the atoms themselves, in the protons, neutrons, and electrons, that suggests they will spontaneously organize themselves into bacteria, or dinosaurs, or primates, or people. No physicist or chemist has any theory from which dinosaurs follow by "selectivity intrinsic to the structures" (p. 31), nor do biologists have any theories by which dinosaurs follow from primeval bacteria. The life story is not coded into the energetic materials. No, the history of life is much more contingent than that. If we were to run the tape all over again, something quite different would happen. From the beginning, the possibilities are there (provided for in our metaphysics by positing "subjectivity" at the foundational level), but what will the actualities be that unfold?

We cannot believe that the long-standing success of the life forms over the millennia is nothing but a contest between organismic individuals, each doing its own thing and bumping into one another by accident. Such a spontaneous, blooming, buzzing confusion would be a pluralist chaos. The only adequate metaphysical hypothesis for all that has occurred is a divine lure. Birch's conclusion finds an ultimate more from which the less comes to be more. That may be too metaphysical, too religious for some. But down-to-Earth explanations do, in the end, have to face up to considerable mystery.

Explanations are not over, but we have made some progress. Humans are biological animals, but they are the only kind of animal that can do metaphysics. That itself is proof of something more that has come out of less. Completing the explanation might take us, as Birch suggests, well on into the next millennium.

References

Eigen, Manfred. 1971. "Selforganization of matter and the evolution of biological macromolecules." *Die Naturwissenschaften* 58:465–523.

Gould, Stephen Jay. 1989. *Wonderful Life: The Burgess Shale and the Nature of History*. New York: W. W. Norton.

Monod, Jacques. 1972. *Chance and Necessity*. New York: Random House.

Leslie, John. 1989. *Universes*. London: Routledge.

Raup, David M. 1991. *Extinction: Bad Genes or Bad Luck?* New York: W. W. Norton.

Wald, George. 1974. "Fitness in the universe: Choices and necessities." In J. Oró, S. L. Miller, C. Ponnamperuma, and R. S. Young, eds., *Cosmochemical Evolution and the Origins of Life* (vol. 1, pp. 7–27). Dordrecht, Holland: D. Reidel Publishing Co.

Glossary

adaptationism   The theory that all the important characteristics of organisms—anatomy, function, and behavior—are to be explained as resulting from natural selection pressures for an adapted fit. Each such feature is present because it has survival value.

altruism   Unselfish concern for the welfare of others, commonly contrasted with egoism, concern for oneself. Many ethicists claim that ethics requires moral altruism, "loving one's neighbor as one does oneself," and how the two combine is the subject of philosophical debate. Biologists, especially sociobiologists, may claim that all apparent ethical altruism is, upon further analysis, actually action in biological self-interest, which may require cooperative actions benefiting others as one benefits oneself.

amino acids   Basic organic molecules that serve as the building blocks of proteins.

ATP   Adenosine triphosphate, an energy molecule used to drive reactions in cells.

autocatalysis   The process by which a molecule catalyzes itself, or facilitates its own self-assembly. RNA may have been, at the origin of life, an autocatalytic molecule.

catalyst   A substance that causes or increases the rate of a chemical change (catalysis), without itself being consumed by that change. Catalytic molecules are essential for many biochemical processes; many catalysts (enzymes) are proteins that the body makes and that are coded on DNA molecules.

catastrophism   In evolutionary history, times of major upset in the history of life have had a dominant effect on the subsequent history of life. Five or six such catastrophic events are frequently recognized. A large meteor crash may have disrupted life at the end of the Cretaceous Period. At other times, there may have been sudden climatic changes. See Eldredge, Chapter 3.

chloroplasts   Organelles in plants that absorb energy from the sun and use it to drive the synthesis of organic materials from carbon dioxide and water.

DNA   Deoxyribonucleic acid, a double-stranded helical molecule, contained in genes, capable of replicating and coding for the inherited structure of the diverse kinds of proteins involved in the metabolisms of life, also coding for their functioning and regulation.

dualism   Any of various philosophical theories that find two fundamental kinds of thing, or ways of being, in the world, as mind and matter, typically with a strong contrast between the two.

egoism   Concern for oneself, contrasted with altruism, concern for others.

entelechy   From a Greek root, *enteleche*, a vital force urging an organism toward self-fulfillment.

enzyme   A protein that serves as a catalyst.

epigenetic rules   Rules of behavior, postulated by sociobiologists, that develop as a human individual grows and matures. These have their ultimate origin in genetic determinants and needs but leave open more flexible behavior, depending on circumstances, education, and cultural opportunities, always with the ultimate objective of leaving more offspring in the next generation. They involve inclinations and dispositions more than genetically stereotyped behavior.

eucaryote   An organism with cells that have nuclei that enclose the genetic material. This includes the all higher organisms, such as protists, fungi, plants, and animals.

Gaia hypothesis   A theory proposed by James Lovelock and advocated by Lynn Margulis and Dorion Sagan that the biotic community of life on earth is best thought of as a living superorganism, to which the name Gaia is given. The temperature and composition of Earth's atmosphere are actively regulated by the sum of life on the Earth, the biota, resulting in the continued global habitability of Earth. Life on Earth does not simply accept geophysical nature and adapt to it, but life interacts to produce environmental conditions that are more favorable to life.

genotype   The genetic makeup of an organism, coded in the genes.

gradualism   In evolutionary theory, the theory that natural selection occurs by gradual, incremental changes, contrasted with more sudden changes, such as those introduced by catastrophes (catastrophism) or by more irregular, steplike changes (punctuated equilibrium).

hominid   Members of the human family. A primate of the family *Hominidae*, which by some classifications includes both the great apes and *Homo sapiens*. Also used to refer to the human species and those fossil species of our direct lineage. If the great apes and *Homo sapiens* are classified in a superfamily, *Hominoidea*, the corresponding term is *hominoid*.

humanum   A Latinized word used to include all the characteristics of human history, especially those that distinguish humans from biological natural history. The reference is to what is unique about humans, such as their self-reflective rational, religious, ethical, political, literary, and similar capacities, transcending our animal nature.

memes   Units of social inheritance, a term proposed, analogously with gene in biological inheritance, for cultural ideas, behaviors, and practices that reproduce themselves and spread through a population, displacing other ideas, behaviors, and practices less successful at reproducing themselves.

metaethics   The study of the foundations of ethics, contrasted with inquiry into what humans ought to do, normative ethics.

mitochondria   Organelles in cells that serve as the sites for respiration. Mitochondria convert energy to forms that the cells can use for various kinds of work of synthesis or metabolism.

molecular fossils   Structures or processes that are widespread in organisms living today and that are thought to reflect structures and processes that remain from the earliest life. See Cech, Chapter 1.

monomer   The subunits that form the building blocks of a polymer. Monomer means "one part."

naturalistic fallacy   An alleged fallacy in ethical argument that moves from what *is* the case, facts about the world or humans, to what *ought* to be the case. A frequent claim is that any argument from biology to ethics commits the naturalistic fallacy.

neoteny   A developmental condition retaining juvenile characteristics in the mature adult stage. Also, the development of adult features in the juvenile condition.

normative ethics   Ethics that asks what humans ought to do, what the norms or standards of human behavior are. Sometimes contrasted with metaethics, the study of the foundations of ethics.

nucleic acid   RNA or DNA, genetic materials found typically in the nucleus of cells and allowing organisms to reproduce as well as to construct life structures and maintain life processes.

nucleoside   A structural unit of DNA. One of several bases (adenine, guanine, cytosine, uracil, thymine) linked with a pentose sugar (ribose) to form a nucleoside. These will in turn form nucleotides.

nucleotide   A structural unit of DNA. A nucleoside linked with a phosphate group is a nucleotide, and these in turn are further linked to form the DNA or RNA chain.

oligonucleotide   A short chain of a few nucleotides, much shorter than DNA or RNA chains.

phenotype   The expressed traits of an organism, resulting as the genotype develops under environmental possibilities, contingencies, and constraints.

pleiotropy   A genetic effect where one gene determines, controls, or has effects on several different organs, anatomical traits, or behaviors.

polymer   A large molecule consisting of many identical or similar parts (monomers) linked together. Polymer means "many parts."

polypeptide chain   A polymer or chain of many amino acids, linked by peptide bonds, that, when appropriately folded and cross-linked, becomes a protein molecule.

procaryote   An organism that does not have a membrane-enclosed nucleus. Procaryotes are found only in the bacteria and cyanobacteria, or blue-green algae; all higher forms of life are eucaryotes.

psyche   The human psychological experience; the capacity to be a person who experiences with a sense of self or ego; loosely, the human mind, soul, or spirit.

punctuated equilibrium   In evolutionary theory, the theory that changes occur by irregular, steplike innovations, often with long periods of relatively little change between such punctuated changes. Gradual and regular incremental changes are less important.

purine   A six-membered ring bound to a five-membered ring of carbon and nitrogen atoms, a base, that is a fundamental building block of DNA and RNA. Two are important: adenine and guanine.

pyrimidine   A six-membered ring of carbon and nitrogen atoms, a base, that is a fundamental building block of DNA and RNA. Three are important: cytosine, thymine, and uracil.

reductionism   Explaining some higher phenomenon in terms of its simpler parts, which are organized so as to produce the higher phenomenon. For example, biological events may be explained in terms of their chemistries or ethics may be explained in terms of its genetic survival benefits (as in sociobiology). There are different kinds of reduction. Some thinkers favor reductionist explanations; others complain that organized wholes display characteristics that exceed the characteristics of the parts.

ribose   A pentose (five-carbon) sugar that is used in constructing DNA and RNA.

ribozyme   RNA or an RNA-like molecule that functions as an enzyme. In this case, a molecule coding information is also functioning as a catalyst, with two important life functions being combined in one molecule. See Cech, Chapter 1.

RNA   Ribonucleic acid, a single-stranded nucleic acid molecule involved in protein synthesis. The structure or coding of RNA is specified by that of DNA and is in turn used to construct proteins. RNA is, in contemporary organisms, an intermediate molecule, but may have once been the primitive life molecule, serving all the functions of life. There are three types of RNA. See Cech. Chapter 1.

RNA polymerase   A protein enzyme that catalyzes the elongation of RNA chains.

soma   The body (from the Greek *soma*, body), especially as this is complemented by or contrasted with the mind, the psyche, or the soul.

superorganism   A biological level of organization higher than the organism yet modeled after organisms. Ecosystems, species, and the planet Earth have been proposed as superorganisms. Sagan and Margulis, following James Lovelock, consider the biota on Earth to be a superorganism, to which they give the name Gaia. Others think that some social organizations, such as nations or churches, are superorganisms, with persons serving roles as parts in the larger whole.

sociobiology   The systematic study of the biological basis of all social behavior, both animal and human. Sociobiologists maintain that all behavior, including human cultural and ethical behavior, has its principal determinants in genes, resulting from natural selection for survival of those who leave the most offspring in the next generation.

social Darwinism   Darwinian evolutionary thought applied to human society. This may include the claim that in human society, too, the fittest survive, with what it means to be fit variously interpreted (sometimes emphasizing combative, sometimes cooperative, behavior). It may includes the claim that there is a gradual evolution toward higher social forms, progress in social development over the centuries.

symbiosis   A relationship where two or more organisms live so that each supplies a need of the other.

Index